Smart Innovation, Systems and Technologies

Volume 228

The Smart Innovation, Systems and Technologies book series encompasses the topics of knowledge, intelligence, innovation and sustainability. The aim of the series is to make available a platform for the publication of books on all aspects of single and multi-disciplinary research on these themes in order to make the latest results available in a readily-accessible form. Volumes on interdisciplinary research combining two or more of these areas is particularly sought.

The series covers systems and paradigms that employ knowledge and intelligence in a broad sense. Its scope is systems having embedded knowledge and intelligence, which may be applied to the solution of world problems in industry, the environment and the community. It also focusses on the knowledge-transfer methodologies and innovation strategies employed to make this happen effectively. The combination of intelligent systems tools and a broad range of applications introduces a need for a synergy of disciplines from science, technology, business and the humanities. The series will include conference proceedings, edited collections, monographs, handbooks, reference books, and other relevant types of book in areas of science and technology where smart systems and technologies can offer innovative solutions.

High quality content is an essential feature for all book proposals accepted for the series. It is expected that editors of all accepted volumes will ensure that contributions are subjected to an appropriate level of reviewing process and adhere to KES quality principles.

Indexed by SCOPUS, EI Compendex, INSPEC, WTI Frankfurt eG, zbMATH, Japanese Science and Technology Agency (JST), SCImago, DBLP.

All books published in the series are submitted for consideration in Web of Science.

More information about this series at http://www.springer.com/series/8767

Suman Bhattacharjee · Siuli Roy · Sipra Das Bit

Post-disaster Navigation and Allied Services over Opportunistic Networks

Springer

Suman Bhattacharjee
Information Technology
Dr. B. C. Roy Engineering College
Durgapur, West Bengal, India

Siuli Roy
Information Technology
Heritage Institute of Technology
Kolkata, West Bengal, India

Sipra Das Bit
Computer Science and Technology
Indian Institute of Engineering Science
and Technology, Shibpur
Howrah, West Bengal, India

ISSN 2190-3018 ISSN 2190-3026 (electronic)
Smart Innovation, Systems and Technologies
ISBN 978-981-16-1239-8 ISBN 978-981-16-1240-4 (eBook)
https://doi.org/10.1007/978-981-16-1240-4

This Springer imprint is published by the registered company Springer Nature Singapore Pte Ltd.
The registered company address is: 152 Beach Road, #21-01/04 Gateway East, Singapore 189721, Singapore

Parents teach in the toughest school in the world—The School for Making People. You are the board of education, the principal, the classroom teacher, and the janitor.

—Virginia Satir

Dedicated to our parents

Foreword by Prof. Sukumar Ghosh

Disasters of various types and scales frequently occur, and they torment human society in unpredictable ways. Each disaster has a unique footprint. While Covid-19 pandemic is once in a century event, the more common disasters are natural disasters triggered by earthquake, flood, tornado etc. The challenge is to maintain emergency services and supply lines so as to minimize the loss of life and property as much as possible. In today's digital society, a key component of all the essential services is communication. The lack of effective communication paralyzes essential services. Road networks become partially inaccessible. Our modern society takes phone and internet services for granted, but often they are the first ones to fall apart when disasters strike. Social media becomes unreachable. The first responders are adversely affected. Adequate preparedness for such disasters requires both proactive efforts and reactive steps.

This monograph proposes and documents a number of strategies to mitigate the impact of a disaster that disrupts electronic communication. These are based on extensive research conducted by the authors over the past few years. Maintaining electronic communication in the face of disasters has been emphasized. Of particular interest here is the study of various types of opportunistic and delay-tolerant networks and their performances in simulated environments. Both practitioners and networking researchers engaged in disaster mitigation will find them useful.

December 2020

Sukumar Ghosh
Professor Emeritus
Department of Computer Science
University of Iowa
Iowa City, IA 52242, USA

Foreword by Prof. Keiichi Yasumoto

The monograph on Post-disaster Navigation and Allied Services over Opportunistic Networks addresses an important and timely topic. In recent years, natural disasters such as earthquakes, tsunamis, landslides, volcanoes, wildfires, floods, hurricanes, tornadoes, etc. have occurred more frequently than ever. Within the recent 50 years from 1967 to 2016, about 8,000 natural disasters occurred in the world and caused huge loss of human (280 million deaths) and economic resources (about 730 billion dollars). The top three regions where natural disasters most frequently happened are South-East Asia, South Asia and Central America. In terms of devastation, in September 2018, Sulawesi earthquake and tsunami happened in Indonesia where more than 4,000 people were killed. In April 2015, the Nepal Earthquake killed nearly 9,000 people, in excess of 600,000 houses were destroyed and more than 288,255 houses were damaged. Due to global warming, the number of natural disasters caused by climate changes such as typhoon, hurricanes and floods is also growing worldwide.

In the aftermath of large-scale disasters, one of the biggest problems is the impairment of cellular communication and the unavailability of internet services caused by the failure of communication infrastructure, including cellular towers and base stations by the power outage, water leakage, etc. Therefore, rapid deployment of post-disaster communication infrastructure and provisioning of internet services, especially GIS service, are crucial. Without communication infrastructure or GIS service, it becomes extremely difficult to facilitate life-saving rescue operations as well as adequate relief services in a timely manner. It is important to establish reliable and seamless information exchange among all stakeholders—disaster victims, emergency and disaster management personnel, policemen, rescue and medical teams, volunteers, government and non-government organizations, etc. This monograph addresses some of these challenges by proposing a post-disaster map-based navigation service, post-disaster situational awareness and resource management, and ad hoc communication infrastructure based on opportunistic networks for realizing these challenges during disaster situations where communication infrastructure is destroyed.

The monograph is well exploring relevant issues and challenges related to post-disaster services in six chapters. Chapter 1 introduces disaster management cycle, post-disaster management services and challenges in post-disaster scenarios followed by architecture, routing and applications of opportunistic networks. Chapter 2 describes the author's primary challenge, post-disaster map-based navigation service, while Chap. 3 deals with the two other important challenges, post-disaster situational awareness and resource management. Chapter 4 presents the opportunistic networks architecture for reliable and energy-efficient post-disaster management. Chapter 5 investigates the routing protocols for energy-efficient post-disaster management. The final chapter summarizes the future research scope in various technologies applied to post-disaster management. The proposed map-based navigation services, situational awareness and resource management services, opportunistic network architecture and routing protocols may help enhance emergency preparedness and personnel training. Thus the monograph is expected to benefit not only researchers and academicians but also disaster management practitioners of government and non-government organizations, particularly in developing countries.

January 2021

Keiichi Yasumoto
Professor of Graduate School of Science and Technology
Nara Institute of Science and Technology
Ikoma, Japan

Preface

It was a nice afternoon of 25 April, 2015, and we were attending a meeting of our research team at an Engineering institute campus away from Kolkata, India. The topic of discussion in that meeting was the application of ICT in disaster management. Suddenly we felt a mild tremor and within a few hours, we got to know that the tremor was caused by a massive earthquake which hit Nepal. The severity of the disaster was enormous, and the whole country was partially disconnected from the rest of the world. After almost one and half months of the disaster, a small team of our research group visited Katmandu, Nepal. The visuals were horrifying at Katmandu, and the road conditions were terrible even after one and half months of the disaster. We observed uncleared debris was piled up at the middle of the roads. There was scarcity of food, drinking water and shelters. Those visuals have motivated us to conduct research so that the disaster-prone countries can equip themselves with appropriate measures to cope with the challenges associated with the disasters. Some of the serious challenges are incapacitated communication infrastructure, unstable power supply and inaccessible road networks. Out of these challenges, the destruction of road networks, especially in developing countries, acts as a major hindrance to effective disaster management. This monograph provides the details of developing a digital route map construction system over the opportunistic networks. The monograph also provides the solution of the two other important post-disaster management services such as situational awareness and resource allocation. Both of these services are invariably dependent on the existence of navigation support. Finally, in order to offer such services, the other challenge is to address the problem of incapacitated communication infrastructure. This monograph also deals with such challenges in post-disaster scenarios and develops automated post-disaster management services.

The monograph is primarily divided into two parts, such as developing post-disaster management services and designing a post-disaster opportunistic network framework underneath. The first three chapters of the monograph discuss the development aspects of numerous post-disaster management services, whereas Chaps. 4 and 5 deal with designing a post-disaster opportunistic network framework.

Chapter 1 of this monograph deals with the necessary background to understand the remaining chapters. Initially, this chapter provides an introduction to the overall disaster management process. The various post-disaster management services along with the challenges of disaster management are discussed in the later parts of the chapter. An overview of opportunistic networks including their features, architecture, routing protocols and applications is also discussed in this chapter.

A smartphone-based digital route map construction system (Post-disaster Map Builder) is discussed in Chap. 2 which offers map-based navigation service in post-disaster scenarios. The proposed system constructs digital route maps of the affected areas using mobile handheld devices such as smartphones.

In Chap. 3, two post-disaster management services, namely situational awareness and resource management are discussed. An Android application (Disaster-Messenger) is presented which allows the sharing of situational information (text, image, video, or voice recordings) without the help of conventional network infrastructure. The development of a decentralized scheme (DPDRM) is also presented in this chapter to automate post-disaster resource management. The proposed scheme allows periodic group-based sharing of resource inventories and supports the search and retrieval of queries within a restricted time period.

In Chap. 4, a reliable and energy-efficient post-disaster opportunistic network architecture is designed to support post-disaster communication. The proposed opportunistic network architecture includes suitable network architecture and node deployment plans. Thus, three-tier architecture is proposed to facilitate post-disaster communication. Suitable node deployment plans for both mobile and static nodes corresponding to tier-1 and tier-2, respectively, to enhance the network reliability are also described in Chap. 4.

Finally, three energy-efficient post-disaster opportunistic network routing protocols, viz., Priority-enhanced PRoPHET (Pen-PRoPHET), Group encounter-based PRoPHET (Ge-PRoPHET) and Trace Route supporting three different post-disaster management services, such as situational awareness, resource management and map-based navigation, respectively, are described in Chap. 5. The aforesaid routing protocols along with the proposed post-disaster communication architecture satisfy the desired network reliability requirement for post-disaster communication.

The primary targeted readers of the monograph are technologists, disaster management agencies, practitioners and researchers working in the domain of ICT-based disaster management in developing countries. They would find it useful in understanding the challenges of providing low-cost post-disaster management services in the disaster aftermath in absence of traditional communication infrastructure. In addition to that, due to the sharp rise in the number of disasters in the recent past, academic institutions are introducing specialized courses on disaster management to enhance student awareness. This monograph can also be used as a reference book in such courses.

Kolkata, India Suman Bhattacharjee
Kolkata, India Siuli Roy
Howrah, India Sipra Das Bit
December 2020

Acknowledgments

Writing a monograph is a project by itself, and just like any real-life project, publication of a monograph requires active cooperation from a large group of people apart from the authors. We would like to thank those people who are involved in the preparation of the manuscript and without whom the monograph could not have seen the light of the day.

First of all, we express our gratitude to Prof. Sukumar Ghosh, Department of Computer Science, University of Iowa, USA, for mentoring our research. His guidance remains influential in the preparation of this monograph.

We are thankful to Prof. Somprakash Bandyopadhyay, MIS Group, Indian Institute of Management Calcutta, India for sharing his thoughts that helped us conceive many works that have been presented in this monograph.

We are indebted to the Society for Promotion of Appropriate Development Efforts (Spade) a voluntary NGO in northeast India for supporting us in different field trials which are crucial for our research. In addition to that, we would like to thank *Doctors For You* (a PAN India NGO) for offering their wholehearted support for acquiring real post-disaster data from various affected areas which are instrumental for our research.

This monograph would not have been completed without the continuous support from the editorial and publishing teams of Springer Nature. In particular, we would like to thank Mr. Aninda Bose and Ms. Sharmila Mary Paneer Selvam for their support from behind the scene.

Finally, we express our gratitude to our family members who wait patiently for the completion of this monograph. Without their affection and continuous support, it would not have been possible for us to complete this monograph.

Kolkata, India Suman Bhattacharjee
Kolkata, India Siuli Roy
Howrah, India Sipra Das Bit
December 2020

Contents

About the Authors

Suman Bhattacharjee is Associate Professor and Head of the Department of Information Technology, Dr. B. C. Roy Engineering College, Durgapur, India. He has received his Ph.D. in Engineering from Indian Institute of Engineering Science and Technology, Shibpur. Prior to that, he has received M.Tech. and B.Tech. degrees from West Bengal University of Technology. Suman has around 10 years of experience in academia, research and software industry. He has got 11 publications till date in international journals and conference proceedings in the area of wireless ad hoc networks and peer-to-peer networks. His current research interest includes opportunistic networks, device-to-device communication and complex networks analysis. He is a professional member of ACM.

Siuli Roy is Professor and Head of the Department of Information Technology, Heritage Institute of Technology, Kolkata, India. She received her Ph.D. in Computer Science Engineering and MCA both from Jadavpur University. She has around 25 years of experience in academic research, software development and business in organizations of international repute including Advanced Telecommunications Research Institute, Japan. Roy has profound research experience with more than 40 research publications in international journals and conference proceedings in the area of wireless networking and pervasive computing. She co-authored two books, namely "Enhancing Performance of Ad Hoc Wireless Networks with Smart antennas", with CRC Press (2005), and "Reliable Post Disaster Services over Smartphone based DTN", with Springer Nature (2019). Roy handled five Govt. funded R&D projects as Principal Investigator/ Co-PI/Technology Partner. Her current interest focuses on empowering the rural artisan community of India with an ICT-enabled capability framework.

Sipra Das Bit is Professor of the Department of Computer Science and Technology, Indian Institute of Engineering Science and Technology, Shibpur, India. She also served as Visiting Professor in the Department of Information and Communication Technology, Asian Institute of Technology, Bangkok, in 2017. A recipient of the Career Award for Young Teachers from the All India Council of Technical Education (AICTE), she has more than 30 years of teaching and research experience. Professor

Das Bit has published many research papers in reputed journals and refereed international conference proceedings. She also has three books to her credit. Her current research interests include Internet of things, wireless sensor network, delay-tolerant network, mobile computing and network security. She is a senior member of IEEE.

Abbreviations

AFW	Autonomous Flight Wireless Nodes
AODV	Ad hoc On-Demand Distant Vector
CAMR	Context-Aware Multicast Routing
DDDAS	Dynamic Data-Driven Application Systems
DM	Data Mules
DPD	Discover–Predict–Deliver
DSR	Dynamic Source Routing
DTN	Delay Tolerant Networks
EASDM	Energy-Aware Single-Data Multicast
GPS	Global Positioning System
ICMN	Intermittently Connected Mobile Network
ICT	Information and Communication Technology
MANET	Mobile Ad hoc Networks
MAP	Mobile Access Point
MCS	Master Control Station
NBC	Naïve Bayesian Classifier
NLOS	Non-line-of-sight
NLP	Natural Language Processing
OMN	Opportunistic Mobile Networks
PRoPHET	Probabilistic Routing Protocol using History of Encounters and Transitivity
PSN	Pocket Switched Networks
RAPID	Resource Allocation Protocol for Intentional DTN
RMS	Root Mean Square
RPLM	Relay Placement Strategy for Latency Minimization
RW	Relief Workers
SDM	Single-Data Multicast
SnW	Spray and Wait
SP	Shelter Point
TB	Throwbox
UAV	Unmanned Arial Vehicles
WIPER	Wireless Phone-based Emergency Response System

Chapter 1
Introduction

Over the last couple of decades, countries across the globe, both developing and developed, have witnessed a significant number of disasters. A disaster is an unanticipated disastrous event that severely disturbs the normal functionality of a community or society and causes heavy casualties as well as enormous damage to property. Typically the entire region is not affected by the disasters uniformly. A few portions are harshly hit owing to their topography. Such portions are called affected areas in disaster management vocabulary [1–3]. Disaster management refers to organizing and managing the resources and liabilities in order to mitigate the influence of disasters. It is further subdivided into two: pre-disaster management and post-disaster management [4]. Pre-disaster management involves the works done in advance of disasters to reduce the severity through prevention and mitigation, as well as improve response through preparedness and planning. Post-disaster management involves the works done in disaster aftermath to reduce human casualties and restore normal social activities in the affected areas with the help of response and recovery. In a post-disaster scenario, normal human activities are severely impaired. As a result, different disaster response agencies mobilize their resources such as human resources, aid, etc. to the affected areas in order to expedite the process of disaster management.

1.1 Disaster Management Cycle

The disaster management cycle demonstrates the continuing practices by which the governments, disaster response agencies and civil society plan for disaster response operations and react after a disaster has occurred. Figure 1.1 illustrates the modern disaster management cycle involving four phases, such as mitigation, preparedness, response and recovery [5–7]. The aforesaid four phases of disaster management are defined by Coppola [8] as follows.

S. Bhattacharjee et al., *Post-disaster Navigation and Allied Services over Opportunistic Networks*, Smart Innovation, Systems and Technologies 228,
https://doi.org/10.1007/978-981-16-1240-4_1

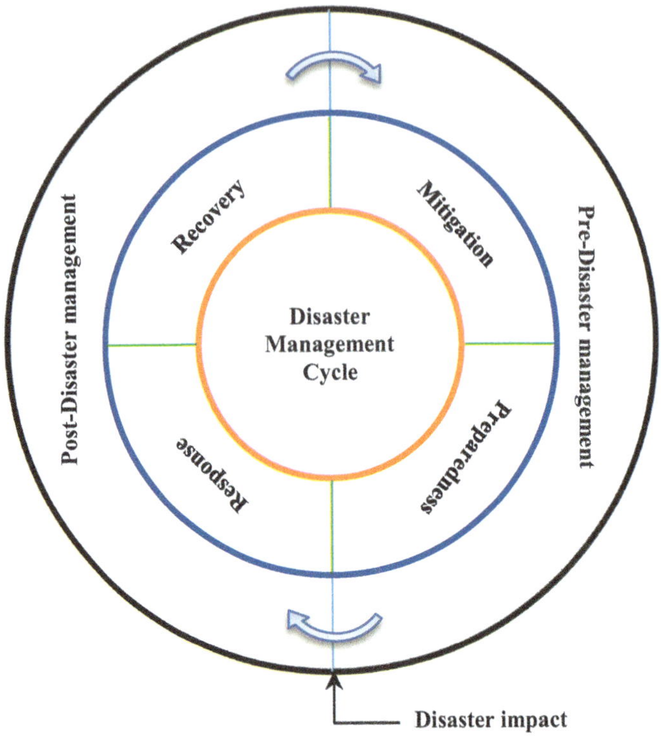

Fig. 1.1 Disaster management cycle

1.1.1 *Mitigation*

Mitigation is a proactive phase aimed at reducing the number of casualties as well as the loss of properties by diminishing the impacts of disasters. This phase reduces or eliminates the impacts and consequences of a disaster through proactive and pre-determined measures, which include an assessment of hazards and recurrent problems, building a plan to mitigate such problems and adopting subsequent actions to implement the plan. The mitigation phase primarily involves three activities, such as risk analysis, risk reduction and risk insurance. The mitigation measures help the community to prevent severe damages to their assets and help them remain operational in the face of disasters.

1.1.2 *Preparedness*

Disaster preparedness typically refers to the actions taken in the pre-disaster scenario in order to ensure adequate response and relief in a post-disaster scenario. This phase involves equipping citizens or the disaster response agencies with adequate tools and knowledge for disaster management so that the human casualties and damage of properties can be minimized in a post-disaster scenario. Numerous establishments and individuals comprising disaster response agencies, government officials, NGOs and common people participate in the preparedness phase. Each has to fulfill a pre-defined responsibility in disaster aftermath. The primary goals of the disaster preparedness phase consist of knowing what to do when disaster strikes, how to do and being prepared with the appropriate tools and techniques to act efficiently. Preparedness minimizes the adverse effects of a disaster by pre-emptive actions that ensure a timely and suitable response as well as the efficient organization and distribution of relief materials.

1.1.3 Response

The response phase is initiated immediately after the onset of the disaster. It includes taking suitable action to reduce the impact of disasters immediately after its occurrence. This is the most complex phase of disaster management conducted in a time-constrained environment with inadequate information regarding the impact of disasters. The primary objective of this phase is saving lives. In addition to that, resource management in terms of the distribution of relief materials like water, food, clothes, medicines, etc. is necessary in order to ensure an adequate supply of relief materials among victims. Coordination is a crucial requirement of the disaster response phase owing to the presence of numerous response agencies in the affected areas. The quality of response depends on adequate information regarding post-disaster scenarios and coordination among different disaster response agencies. In addition to that, the response phase requires immediate recommencement of critical infrastructure such as opening transportation routes, restoring communication, electricity and water supply in order to minimize further human casualties and expedite the process of the normal functioning of society.

1.1.4 Recovery

The recovery phase starts immediately after the disaster is over and generally extends up to several weeks or even months depending on the severity of the disaster. The main objective of this phase is social continuity management which implies the restoration of usual day-to-day operations in the affected areas. The actions related to the recovery phase are the most diversified in nature. The recovery phase generally consists of activities such as restoration of emergency services (e.g. communication, water supply, power supply), disaster impact assessment and rehabilitation [9–11].

The aforesaid discussion clearly reveals that out of the four phases of disaster management the mitigation and preparedness phases are associated with pre-disaster management, whereas the response and recovery phases are related to post-disaster management. A set of disaster management services is associated with each phase. However, after the onset of disaster the post-disaster management is crucial in order to minimize unwanted casualties and damage to properties. Hence, in the following section, we briefly discuss post-disaster management services.

1.2 Post-disaster Management Services

As per the Disaster Assistance Guide to Recovery Programs, FEMA [12], National Post-Disaster Recovery Planning and Coordination, UNDP [13] and Parasuraman et al. [14], all the activities associated with disaster response and recovery phases as discussed in Sect. 1.1 can be broadly categorized into six post-disaster management services such as map-based navigation [15–18], situational awareness [19, 20], resource management [21–23], healthcare [24, 25], logistics [26–28] and reconstruction [29–32]. The aforesaid services are discussed below.

1.2.1 Map-Based Navigation

Map-based navigation service is one of the crucial requirements for disaster response operations. In the aftermath of massive disasters, the road networks of the affected areas alter significantly. Various prevailing roads remain unapproachable due to the incidental destruction. Thus, the scarcity of navigation support in such scenarios creates difficulties for the volunteers and victims in reaching their targets through unknown routes which may lead to undesirable human casualties. Moreover, the supply vehicles carrying relief materials face the challenge of reaching their destinations without adequate navigation support, causing the disproportionate distribution of relief materials. Hence, the post-disaster response operation is severely impaired in absence of proper navigation service. Such service guides volunteers to reach their destinations, helps victims to find the shelters through unknown routes and supports disaster response agencies in mobilizing their manpower, machineries and aid to the affected areas [16, 17].

1.2.2 Situational Awareness

Situational awareness remains another vital requirement for managing large-scale disasters efficiently. In respect of disaster management situational awareness refers to the knowledge about the post-disaster situation which typically includes the size of the disaster-affected areas, demographic information, topographic information regarding affected areas, the severity of the damage, type of disaster and availability of resources, including both human and material resources (relief materials, vehicles, life-saving instruments, debris clearing tools etc.) [33]. Effective situational awareness consists of three steps, viz., collecting of situational data, identifying critical factors from such data (situational information) and predicting actionable insights in the near future.

The primary objective of situational awareness is to gain accurate information rapidly such that the response efforts become effective. A well-coordinated post-disaster response operation demands an efficient distribution of situational information among response agencies. Such information helps the response agencies to suitably coordinate, manage and channelize their assets [34], as well as in decision making. The lack of prior knowledge regarding the post-disaster situation causes an inconsistent distribution of resources and delay in taking critical decisions which makes post-disaster response efforts ineffective.

1.2.3 Resource Management

Resource management is an integral part of disaster response operations. The primary objective of resource management is to ensure proportionate and adequate distribution of relief materials. In the disaster aftermath, resources are managed in a decentralized manner by different response agencies. The agencies typically establish different types of camps depending on the type of services they offer. The volunteers associated with each camp provide services relevant to that camp. The resources (e.g. medicines, foodstuffs, clothes and other relief materials) which are essential for providing aid to the victims [35] are stored at each camp. These resources are commonly managed by the camps manually using a decentralized manner in the developing countries. Managing resources through the manual process may generate inconsistency in resource inventories and propagate misleading information among the volunteers about the available inventories. Thus, effective resource management is severely hampered due to inaccuracy in the search and retrieval of available resource information from such inconsistent inventories.

1.2.4 Healthcare

Healthcare is another crucial service for disaster management. This service is primarily responsible to reduce human casualties substantially. Disasters cause health hazards and casualties either directly or indirectly, leaving adequate healthcare facilities unreachable to the affected communities during post-disaster scenarios. Empirical evidence reveals that such negative effects are disproportionally focused in the developing countries which accounts for around 68.2% of globally reported disaster mortalities [36]. In developed countries, efficient and effective emergency medical care systems play a pivotal role in post-disaster response operations by successfully managing medical emergencies. However, the primary healthcare system in developing countries generally plays a vital role in establishing the local capacity and making the communities more resilient to disasters.

1.2.5 Logistics

Logistics is a key component in the disaster response operation. It is defined as the "*management of the flow of things between the point of origin and point of consumption*". In the context of disaster response operation logistics refers to the transportation of equipment, volunteers and resources to the disaster-affected areas. Moreover, the removal of debris is another crucial aspect of post-disaster logistics. Efficient logistics service also requires coordination between different disaster response agencies. Well-organized disaster response operation typically relies on the speed of evacuation and transportation of relief materials at the proportionate amount, at the right time to the disaster-affected areas. Hence, in absence of efficient logistics service, post-disaster response operation is severely hampered.

1.2.6 Reconstruction

Reconstruction service is essential in post-disaster response as well as the recovery phase [31]. In the response phase, reconstruction refers to the rapid construction of critical infrastructures such as roads, bridges, communication infrastructures as well as temporary shelters for the victims. In the recovery phase, reconstruction service is associated with the restoration of damaged houses for the victims, schools, hospitals, economic and government establishments, etc. In case of large-scale disasters, reconstruction service in the recovery phase typically last up to years involving diverse agencies. Hence, post-disaster reconstruction requires a significant amount of coordination and planning in order to become successful.

In the following section, we briefly discuss the challenges of post-disaster scenarios which act as a major hindrance to effective disaster management.

1.3 Challenges in Post-disaster Scenarios

The primary challenges experienced by disaster response agencies and victims in post-disaster scenarios can be broadly categorized into three such as inaccessible routes, incapacitated communication infrastructure and unstable power supply [37–39].

1.3.1 Inaccessible Road Networks

Disasters such as floods, landslides, earthquakes, etc. damage existing road networks severely [40]. As a result, many existing routes for reaching disaster-affected areas become inaccessible. Hence, existing relief maps of the affected areas become outdated. Moreover, in digital mapping systems like Google Maps, Bing Maps, etc. updation of route maps are occasional (typically requires one to three years) in non-US regions. Therefore, the route maps offered by the aforesaid digital mapping systems do not reflect the altered routes in the affected areas after disasters. Hence, even if the communication infrastructure is not affected, the appropriateness of such systems for providing navigation assistance to the disaster response agencies remains uncertain. Thus, different disaster response agencies face difficulties in mobilizing relief materials and manpower to the disaster-affected areas through the existing routes. The success of a disaster response operation generally depends on the speed of evacuation and transportation of an adequate amount of relief materials at the right time to the disaster-affected areas. The inaccessibility of existing road networks reduces the speed of post-disaster response operations drastically hampering the efficiency of disaster management. Therefore, developing a post-disaster map-based navigation service is of utmost importance.

1.3.2 Incapacitated Communication Infrastructure

According to the World Disasters Report [41], the absence of a stable communication network becomes the key issue in post-disaster scenarios. Services like cell phone/Internet connectivity become non-functional immediately after massive disasters due to the destruction/exhaustion of the supporting infrastructure [39, 42, 43]. Hence, the possibility of information exchange through traditional communication infrastructure is almost ruled out. The efficiency of post-disaster response operation relies heavily on the success of sharing situational information among the volunteers of different response agencies. Such information can assist to appropriately coordinate and manage the post-disaster response operation. Since many portions of the traditional communication infrastructure may get undermined due to the disaster, accumulation of data from the different parts of the affected areas and, consequently, planning of the disaster response operation become challenging. The incapacitated communication infrastructure causes several challenges in post-disaster scenarios such as lack of situational awareness, ad hoc allocation and distribution of resources, and inefficiency of map-based navigation services.

The term situational awareness refers to the knowledge about the post-disaster situation in the context of disaster management, which typically includes the size of the disaster-affected areas, demographic information, topographic information regarding affected areas, the severity of damage and type of disaster and availability of resources. However, in disaster aftermath, the communication network becomes unstable. Hence, the accumulation of situational information becomes challenging.

In post-disaster scenario resources (e.g. medicines, foodstuff, clothes and other relief materials) are managed (allocated and distributed) manually in a decentralized manner by different response agencies using pens and papers. This manual process along with intermittent network connectivity causes unawareness about the current status of the resource inventories among the volunteers, and consequently, effective resource management is hampered. As a result, the disproportionate distribution of life-saving resources takes place, which causes unwanted human casualty [44].

In the aftermath of the massive disasters, digital mapping systems become inaccessible in the affected areas due to the unstable network connectivity. Hence, even if the road networks remain unaltered, due to the inaccessibility of such web-based digital mapping systems, providing navigation support in the affected areas becomes uncertain.

1.3.3 Unstable Power Supply

Disasters hamper the reliability and operation of conventional and renewable power generation and supply components [45, 46]. For instance, high winds during storms lead to faults and damages to overhead power transmission lines. Floods damage power supply equipment. Earthquakes destroy power plants and power grids. Thus, the power supply remains unstable in the disaster aftermath. In post-disaster scenarios, one of the most crucial infrastructure interruptions is power outages. This is critical because the communication between different disaster response agencies requires communication systems that depend on a steady power supply. Furthermore, public services such as water supply, solid waste removal and healthcare are adversely affected by the power outage. For example, power outages at hospitals in disaster aftermath can prove lethal for people with chronic diseases and severe injuries that require power for life support equipment.

The aforesaid discussion reveals that in order to provide post-disaster navigation and allied services, there is an acute need for establishing a post-disaster communication network as an alternative to conventional communication infrastructure [47]. In recent years, researchers in the networking domain are strongly advocating for the use of opportunistic networks which are intermittently connected device-to-device networks supporting emergency communication in absence of stable conventional communication infrastructure during disaster response operations [48, 49]. Thus, opportunistic networks prove to be a viable alternative for information exchange in the affected areas. Hence, an overview of opportunistic networks is provided in the following section.

1.4 Overview of Opportunistic Networks

The existing TCP/IP-based Internet has been recognized as a suitable communication system in our day-to-day life where end-to-end connectivity among different entities is prevalent. However, the availability of such communication systems is often ruled out in *challenged areas* [48] like outer space, extreme terrains, battlefields, post-disaster scenarios, etc. A communication system such as mobile ad hoc networks (MANET) has been developed [50] to facilitate connectivity in such scenarios. However, frequent node mobility and low node density may cause network partitioning in ad hoc networks which results in the non-existence of an end-to-end path between source and destination. The non-existence of the end-to-end path hinders the ad hoc routing protocols in delivering messages to their destinations. Hence, message drops become inevitable and message delivery to the destinations is severely hampered. Such challenges can be addressed by introducing a distinctive ad hoc network known as the opportunistic networks. The opportunistic network works under a store–carry–forward paradigm where the nodes store and carry messages to the intermediate or destination nodes in absence of the end-to-end path. Figure 1.2 illustrates the store–carry–forward paradigm employed in opportunistic networks. The opportunistic networks literature is very rich in nomenclature [51].

Apart from the name opportunistic networks, it is also known by various terminologies like disruption tolerant networks, delay tolerant networks (DTN), opportunistic mobile networks (OMN), pocket switched networks (PSN) etc. Although the terminologies vary, all the aforesaid networks exhibit identical characteristics. Hence, throughout this monograph, the terms opportunistic networks and DTN are used interchangeably. Opportunistic networks create an overlay architecture on top of any conventional network protocol stack and provide a gateway to facilitate store–carry–forward function when a node moves in the vicinity of two or more networks [48]. Opportunistic networks are characterized by intermittent connectivity, long delays and interoperability among the nodes of heterogeneous networks.

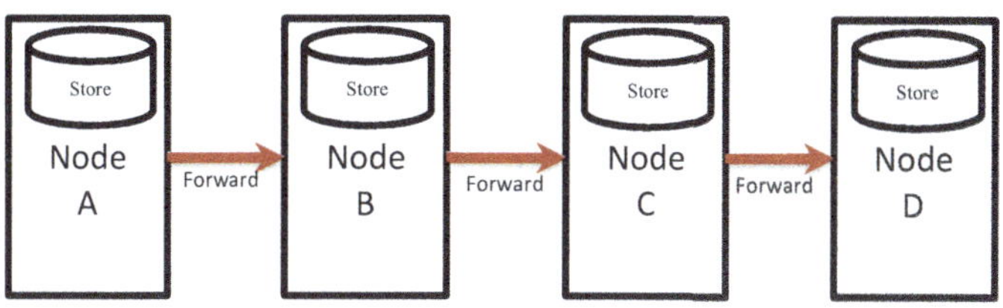

Fig. 1.2 Illustration of the store–carry–forward paradigm

1.4.1 Features of Opportunistic Networks

Opportunistic network possesses a distinctive set of features which are illustrated below:

- **Intermittent connectivity**: As opposed to the conventional communication system, an opportunistic network does not provide stable network connectivity. Due to node mobility, the end-to-end path between source and destination is non-existent. The non-existence of the end-to-end path between source and destination causes frequent network partitioning which results in intermittent connectivity.
- **Long or variable delay**: Another unique characteristic of an opportunistic network is a long or variable delay. The delays in the opportunistic network may range from hours, days and even months. The end-to-end path delays comprise propagation delays among nodes and variable queuing delays at each node. Such a high delay makes the conventional acknowledgment-based Internet protocols non-functional.
- **Asymmetric data rates**: Conventional Internet protocols support a moderate asymmetric data rate. However, intermittent connectivity and long delay result in highly asymmetric data rates in an opportunistic network.
- **High error rates**: The mobility of nodes facilitates data transfer in opportunistic networks. However, the high mobility of nodes causes frequent network disconnection and increases transmission errors in opportunistic networks. High error rates involve retransmission of the data which causes enhanced network traffic.

1.4.2 Architecture

The opportunistic network architecture is primarily designed to handle frequent network disconnection, long delays and interoperability among the nodes of heterogeneous networks. Figure 1.3 illustrates the layered opportunistic network architecture.

The example includes three network regions (X, Y and Z), consisting of different nodes such as opportunistic network hosts and opportunistic network gateways. The opportunistic network gateways are responsible for transmitting bundles between different network regions. The interoperability among the nodes of different networks is facilitated by introducing the opportunistic network gateways at the interconnection points of such networks. An opportunistic network implements the store–carry–forward mechanism by introducing an additional layer known as the bundle layer over the transport layer in order to minimize the effect of long delays and frequent network partitioning. The bundle layer uses persistent storage to store data. In addition to that, the bundle layer also offers an optional hop-by-hop transfer of reliable delivery responsibility, called bundle custody transfer and an optional end-to-end acknowledgment functionality. When nodes accept custody of a bundle (asynchronous message), they retain a copy of the bundle until such responsibility is transferred to another node [48].

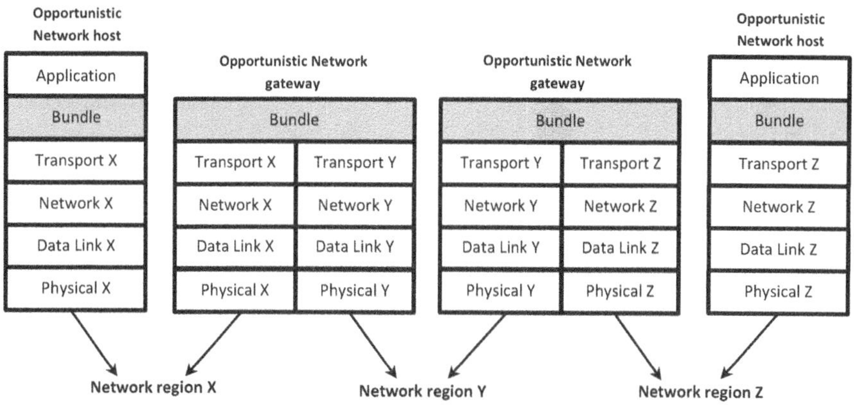

Fig. 1.3 Layered opportunistic network architecture

1.4.3 Routing

Routing is the process of identifying a path for transmitting data within a network or across multiple networks. It is a crucial feature of any communication network including opportunistic networks/DTN. Unlike, conventional infrastructure-oriented networks (wired or wireless) or MANET, the end-to-end path between source and destination is non-existent in opportunistic networks. So, the traditional network routing protocols, such as distant vector routing (DVR), path vector routing (PVR), link state routing (LSR) as well as the MANET routing protocols like ad hoc on-demand distant vector (AODV), dynamic source routing (DSR), etc., prove unsuitable for opportunistic networks. Further, the opportunistic network nodes store and carry messages until a suitable relay node to forward those messages toward the destination is encountered. Due to the intermittent connectivity and long delay, it is challenging to identify a suitable relay node for message forwarding in opportunistic networks.

Thus, an opportunistic network has its own routing protocols, which can be broadly categorized into two, viz., forwarding-based routing (single copy) and replication-based routing (multicopy). The forwarding-based routing protocols require lesser network resources as only a single copy of a message exists at any instant in the network. These protocols abolish the need for acknowledgments produced by the destination in order to eliminate undelivered copies of messages remaining in the network. But, forwarding-based routing protocols are incapable of guaranteeing an adequate message delivery rate in an opportunistic network [52]. On the contrary, a superior message delivery rate can be achieved by implementing replication-based routing protocols because numerous replicas of a message are present in the network, out of which only one needs to reach the destination. The consumption of precious network resources (e.g. bandwidth and energy) remains the primary limitation of replication-based routing protocols [53]. In spite of the inherent shortcomings, the

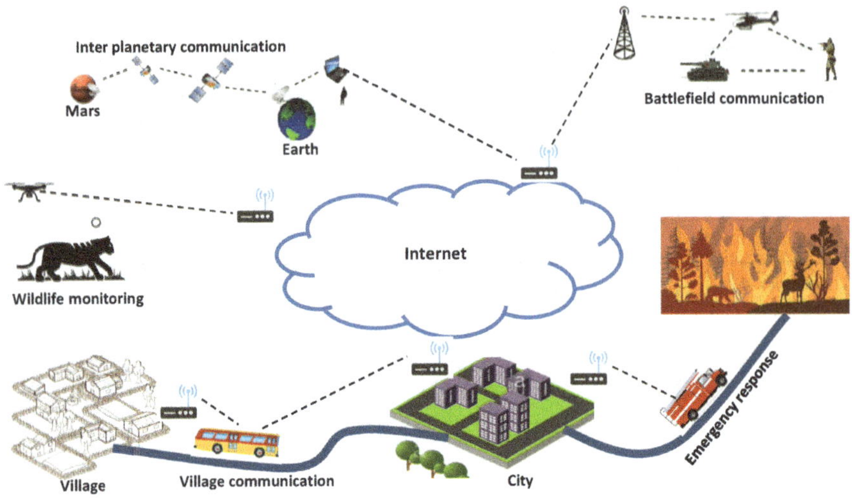

Fig. 1.4 Application of opportunistic networks in *challenged areas*

replication-based routing protocols have been extensively adopted by the opportunistic network research fraternity because of their capability to deliver messages at a higher rate as opposed to the forwarding-based routing protocols.

1.4.4 Applications

As discussed earlier, an opportunistic network can be deployed as a suitable mode of communication in *challenged areas* like outer space, extreme terrains, battlefields and post-disaster scenarios. Figure 1.4 illustrates the applications of opportunistic networks in different *challenged areas*. Several implementations of opportunistic network-based communication networks in *challenged areas* are available till date. Some of the instances of such implementations are *DakNet* [54], *Bytewalla* [55], *DTN MapEx* [17], *pSync* [56, 57] etc.

1.4.4.1 DakNet

DakNet has been developed to offer low-cost communication facilities in rural areas. It consists of kiosks, mobile access points (MAP) and hubs (Internet access points). The MAPs are equipped with portable storage devices and usually mounted on vehicles like buses, motorcycles, bicycles, etc. Hubs are generally located in a nearby town or city where stable Internet connectivity is prevalent. When the vehicles equipped with MAPs come within the proximity of kiosks, data exchange between them takes place. Afterward, when a MAP reaches the nearby hub data from the kiosks are synchronized with the Internet. Thus, the MAPs act as *data mules*.

1.4.4.2 Bytewalla

Bytewalla is an Android application that aims at connecting African rural villages using smartphones-based opportunistic networks. It has the capability of exchanging emails through opportunistic networks. It employs opportunistic network servers which transform emails into bundles and forward them to the smartphones. People visiting from village to city accompanying their smartphones carry the bundles with them to the opportunistic network servers located in the city. The opportunistic network servers at the city extract the emails from the bundle and forward it to the destination through the Internet.

1.4.4.3 DTN MapEx

DTN MapEx is a mapping system that constructs the route maps of the affected areas through an opportunistic network. In order to construct the maps, it employs two different types of nodes, such as sensor and computing nodes. The mobile handheld devices used to collect GPS traces act as sensor nodes. The traces collected by the mobile nodes are subsequently dispatched to the computing nodes (PCs) for the construction of route maps. The route maps constructed by the computing nodes are then conveyed to the sensor nodes over opportunistic networks.

1.4.4.4 pSync

P2P file syncing tool (*pSync*) is a file syncing tool using opportunistic networks. It incorporates numerous cutting-edge features over conventional file syncing tools such as role-based file transfer and device priority scheduling. The role-based file transfer protocol is employed where devices having maximum contact duration are given priority for file transfer. This mechanism provides gives optimum utilization of accessible network resources. In addition to that UDP-based networking is implemented in *pSync* in order to handle intermittent connectivity as well as short contact durations among the devices.

The aforesaid applications of opportunistic networks reveal its potential as a low-cost alternative for the conventional communication network in a post-disaster scenario.

1.5 Motivation

The challenges experienced by disaster response agencies and victims in post-disaster scenarios can be broadly categorized into three, such as inaccessible routes, incapacitated communication infrastructure and unstable power supply. Out of these challenges, the destruction of road networks, especially in developing countries, acts as a major hindrance to effective disaster management. To be more specific, the success of a disaster response operation generally depends on the speed of evacuation and transportation of an adequate amount of relief materials at the right time to the disaster-affected areas. Hence, map-based navigation support is a primary requirement for post-disaster relief operations. The two other important post-disaster management services such as situational awareness and resource allocation are invariably dependent on the existence of navigation support.

In addition to that, in developing countries, post-disaster management services involve manual procedures. For instance, the route maps of the affected areas are drawn manually on pieces of paper and distributed. The situational information is accumulated by filling up forms using the inputs from the affected population. The resource inventories are managed using pens and papers. This manual procedure of post-disaster management consumes a substantial amount of time and prone to human errors resulting in the disproportionate distribution of aids and causing unwanted human fatalities. So the need for automated post-disaster management services is of utmost importance. In order to offer automated post-disaster management services, the principal requirement is to address the challenge of incapacitated communication infrastructure.

It has been observed that conventional communication infrastructure becomes incapacitated in the disaster aftermath due to the incidental damages and destruction. Hence, the existence of stable network connectivity in the affected areas is almost ruled out. These challenges are the motivating factors for developing automated post-disaster navigation, situational awareness and resource management services in absence of conventional communication infrastructure.

1.6 Contribution

The overall contribution of this monograph is offering post-disaster navigation and allied services (situational awareness and resource management) over opportunistic networks (DTN). In order to offer such services, the existence of a post-disaster opportunistic network framework becomes crucial. Therefore, from the broader perspective, this monograph is divided into two major parts such as developing post-disaster management services and designing a post-disaster opportunistic network framework underneath. A post-disaster opportunistic network framework comprises

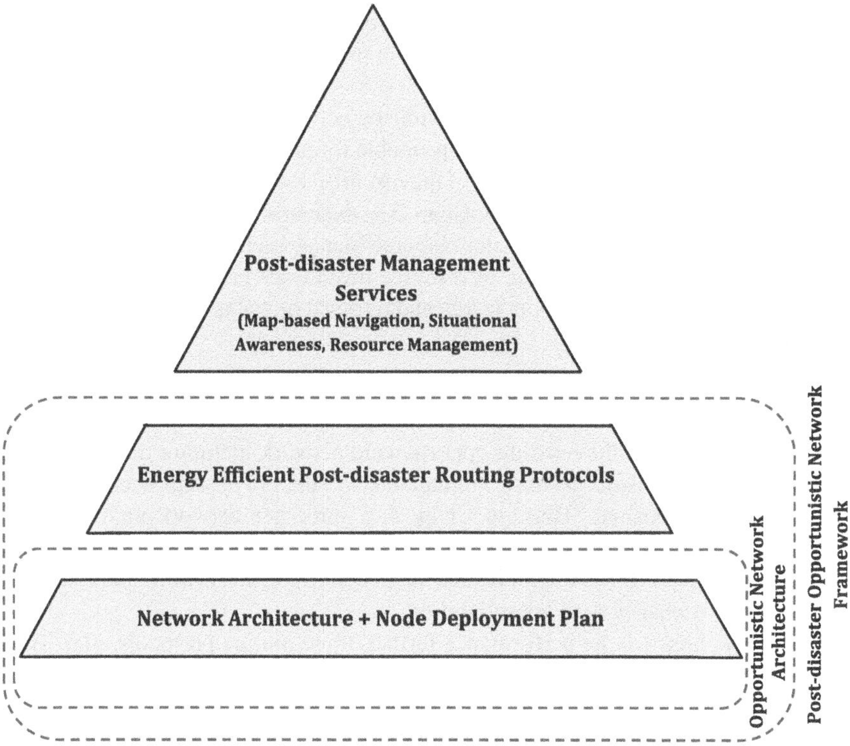

Fig. 1.5 Schematic representation of the opportunistic network framework

two entities such as opportunistic network architecture and energy efficient post-disaster routing protocols. The post-disaster opportunistic network architecture typically includes suitable network architecture and node deployment plans. Figure 1.5 shows a schematic representation of such an opportunistic networks framework.

Developing Post-disaster Management Services:

- There is a scarcity of suitable approaches in constructing route maps from trajectory traces in post-disaster scenarios where high power computing resources are unavailable. Thus, in Chap. 2, a smartphone-based digital pedestrian map construction system (Post-disaster Map Builder) is developed supporting map-based navigation service in post-disaster scenarios. The proposed map-based navigation system accumulates the trajectory traces from the volunteers' handheld mobile devices and constructs a digital map of the disaster-struck areas.
- Situational awareness requires the effective gathering of situational information. However, it is observed from the existing works that the existence of stable network connectivity is indispensable for gathering situational information. In addition to that information gathered from platforms such as social media lacks authenticity. Hence, an Android application (DisasterMessenger) is developed in Chap. 3,

which enables the sharing of situational information (text, image, video or voice recordings) in scenarios where the conventional communication infrastructure is non-existent.

- It is observed from the existing opportunistic network literature that the existing content sharing mechanisms prove unsuitable for post-disaster resource management where detection and delivery of the resource availability information remain strictly time-bound. Therefore, in Chap. 3, a decentralized scheme (DPDRM) is proposed to automate post-disaster resource management. The proposed scheme facilitates group-based sharing of resource inventories periodically and in addition to that offers provision for search and retrieval of resource requests within an acceptable delay.

Designing Post-disaster Opportunistic Network Framework:

- It is observed from the existing opportunistic network literature that an energy-efficient opportunistic network architecture to support post-disaster communication is non-existent. Thus, in Chap. 4, a three-tier network architecture is proposed to facilitate energy-efficient post-disaster communication. Suitable node deployment plans for both mobile and static opportunistic network nodes are also proposed to enhance network reliability.
- In order to improve the performance further, three routing protocols, viz., Trace Route, Priority enhanced PRoPHET (Pen-PRoPHET) and Group encounter-based PRoPHET (Ge-PRoPHET), and supporting three different post-disaster management services such as map-based navigation, situational awareness and resource management, respectively, are proposed in Chap. 5.
- Chapter 6 concludes the monograph with direction for future research.

References

1. Mohamed Shaluf, I.: Disaster types. Disaster Prev. Manag. Int. J. **16**, 704–717 (2007). https://doi.org/10.1108/09653560710837019
2. Schulz, S.F.: Disaster relief logistics: benefits of and impediments to horizontal cooperation between humanitarian organizations. Haupt (2009)
3. Kron, W., Steuer, M., Löw, P., Wirtz, A.: How to deal properly with a natural catastrophe database—analysis of flood losses. Nat. Hazards Earth Syst. Sci. **12**, 535–550 (2012). https://doi.org/10.5194/nhess-12-535-2012
4. Chen, X., Kwan, M.P., Li, Q., Chen, J.: A model for evacuation risk assessment with consideration of pre- and post-disaster factors. Comput. Environ. Urban Syst. **36**, 207–217 (2012). https://doi.org/10.1016/J.COMPENVURBSYS.2011.11.002
5. Akgün, İ., Gümüşbuğa, F., Tansel, B.: Risk based facility location by using fault tree analysis in disaster management. Omega **52**, 168–179 (2015). https://doi.org/10.1016/J.OMEGA.2014.04.003
6. Giannakis, M., Papadopoulos, T.: Supply chain sustainability: a risk management approach. Int. J. Prod. Econ. **171**, 455–470 (2016). https://doi.org/10.1016/J.IJPE.2015.06.032
7. Renn, O.: Risk Governance. Routledge (2017)
8. Coppola, P.D.: Introduction to International Disaster Management (2015)

9. Holguín-Veras, J., Jaller, M., Van Wassenhove, L.N., Pérez, N., Wachtendorf, T.: On the unique features of post-disaster humanitarian logistics. J. Oper. Manag. **30**, 494–506 (2012). https://doi.org/10.1016/J.JOM.2012.08.003
10. Labadie, J.R.: Auditing of post-disaster recovery and reconstruction activities. Disaster Prev. Manag. Int. J. **17**, 575–586 (2008). https://doi.org/10.1108/09653560810918612
11. Fetter, G., Rakes, T.: Incorporating recycling into post-disaster debris disposal. Socioecon. Plan. Sci. **46**, 14–22 (2012). https://doi.org/10.1016/J.SEPS.2011.10.001
12. FEMA: Disaster Assistance A Guide to Recovery Programs, pp. 1–3 (2005)
13. UNDP: National post-disaster recovery planning and coordination. Clim. Change Disaster Risk Reduct. Clust. 1–106 (2016)
14. Parasuraman, S., Unnikrishnan, P.V.: Disaster response in India: an overview. Indian J. Soc. Work **63**, 151–172 (2002)
15. Goodchild, M.F., Glennon, J.A.: Crowdsourcing geographic information for disaster response: a research frontier. Int. J. Digit. Earth **3**, 231–241 (2010). https://doi.org/10.1080/17538941003759255
16. Zook, M., Graham, M., Shelton, T., Gorman, S.: Volunteered geographic information and crowdsourcing disaster relief: a case study of the Haitian earthquake. World Med. Health Policy **2**, 6–32 (2010). https://doi.org/10.2202/1948-4682.1069
17. Trono, E.M., Arakawa, Y., Tamai, M., Yasumoto, K.: DTN MapEx: disaster area mapping through distributed computing over a Delay Tolerant Network. In: 2015 Eighth International Conference on Mobile Computing and Ubiquitous Networking (ICMU), pp. 179–184. IEEE (2015)
18. Lucas, G., Lénárt, C., Solymosi, J.: Development and testing of geo-processing models for the automatic generation of remediation plan and navigation data to use in industrial disaster remediation. Open Geospatial Data Softw. Stand. **1**, 5 (2016). https://doi.org/10.1186/s40965-016-0006-z
19. Seppänen, H., Virrantaus, K.: Shared situational awareness and information quality in disaster management. Saf. Sci. **77**, 112–122 (2015). https://doi.org/10.1016/J.SSCI.2015.03.018
20. Cameron, M.A., Power, R., Robinson, B., Yin, J.: Emergency situation awareness from twitter for crisis management. In: Proceedings of the 21st International Conference Companion on World Wide Web—WWW '12 Companion, p. 695. ACM Press, New York, NY, USA (2012)
21. Muaafa, M., Concho, A.L., Ramirez-Marquez, J.: Emergency resource allocation for disaster response: an evolutionary approach. SSRN Electron. J. (2014). https://doi.org/10.2139/ssrn.2848892
22. Yang, Z., Zhou, H., Gao, X., Liu, S.: Multiobjective model for emergency resources allocation. Math. Probl. Eng. **2013**, 1–6 (2013). https://doi.org/10.1155/2013/538695
23. Chacko, J., Rees, L.P., Zobel, C.W.: Improving resource allocation for disaster operations management in a multi-hazard context. In: ISCRAM 2014 Conference Proceedings—11th International Conference on Information Systems for Crisis Response and Management, pp. 85–89 (2014)
24. Daily, E., Padjen, P., Birnbaum, M.: A review of competencies developed for disaster healthcare providers: limitations of current processes and applicability. Prehosp. Disaster Med. **25**, 387–395 (2010). https://doi.org/10.1017/S1049023X00008438
25. Misra, S., Chatterjee, S.: Social choice considerations in cloud-assisted WBAN architecture for post-disaster healthcare: data aggregation and channelization. Inf. Sci. (Ny) **284**, 95–117 (2014). https://doi.org/10.1016/J.INS.2014.05.010
26. Sheller, M.: The islanding effect: post-disaster mobility systems and humanitarian logistics in Haiti. Cult. Geogr. **20**, 185–204 (2013). https://doi.org/10.1177/1474474012438828
27. Holguín-Veras, J., Pérez, N., Jaller, M., Van Wassenhove, L.N., Aros-Vera, F.: On the appropriate objective function for post-disaster humanitarian logistics models. J. Oper. Manag. **31**, 262–280 (2013). https://doi.org/10.1016/J.JOM.2013.06.002
28. Hu, Z.-H., Sheu, J.-B.: Post-disaster debris reverse logistics management under psychological cost minimization. Transp. Res. Part B Methodol. **55**, 118–141 (2013). https://doi.org/10.1016/J.TRB.2013.05.010

29. Davidson, C.H., Johnson, C., Lizarralde, G., Dikmen, N., Sliwinski, A.: Truths and myths about community participation in post-disaster housing projects. Habitat Int. **31**, 100–115 (2007). https://doi.org/10.1016/J.HABITATINT.2006.08.003
30. Lawther, P.M.: Community involvement in post disaster re-construction—case study of the British RED CROSS Maldives Recovery Program. Int. J. Strateg. Prop. Manag. **13**, 153–169 (2009). https://doi.org/10.3846/1648-715X.2009.13.153-169
31. Le Masurier, J., Rotimi, J.O.B., Wilkinson, S.: Comparison between routine construction and post-disaster reconstruction with case studies from New Zealand (2006)
32. Chang, Y., Wilkinson, S., Potangaroa, R., Seville, E.: Donor-driven resource procurement for post-disaster reconstruction: constraints and actions. Habitat Int. **35**, 199–205 (2011). https://doi.org/10.1016/J.HABITATINT.2010.08.003
33. Mohsin, B., Steinhäusler, F., Madl, P., Kiefel, M.: An innovative system to enhance situational awareness in disaster response: what are end users looking for in such systems. J. Homel. Secur. Emerg. Manag. **13**, 301–327 (2016). https://doi.org/10.1515/jhsem-2015-0079
34. Basu, M., Bandyopadhyay, S., Ghosh, S.: Post disaster situation awareness and decision support through interactive crowdsourcing. Procedia Eng. **159**, 167–173 (2016). https://doi.org/10.1016/j.proeng.2016.08.151
35. Basu, S., Roy, S., Bandyopadhyay, S., Das Bit, S.: A utility driven post disaster emergency resource allocation system using DTN. IEEE Trans. Syst. Man Cybern. Syst. 1–13 (2018). https://doi.org/10.1109/TSMC.2018.2813008
36. Swathi, J.M., González, P.A., Delgado, R.C.: Disaster management and primary health care: implications for medical education. Int. J. Med. Educ. **8**, 414–415 (2017). https://doi.org/10.5116/ijme.5a07.1e1b
37. Pandey, M.: Disaster Management. Wiley, India (2014)
38. Pedroso, F.F., Teo, J., Seville, E., Giovanazzi, S., Vargo, J.: Post-disaster challenges and opportunities: lessons from the 2011 Christchurch earthquake and Great Eastern Japan earthquake and tsunami. In: 2015 Global Assessment Report on Disaster Risk Reduction, pp. 1–27 (2013)
39. Kishorbhai, V.Y., Vasantbhai, N.N.: AON: a survey on emergency communication systems during a catastrophic disaster. Procedia Comput. Sci. **115**, 838–845 (2017). https://doi.org/10.1016/j.procs.2017.09.166
40. Bíl, M., Vodák, R., Kubeček, J., Bílová, M., Sedoník, J.: Evaluating road network damage caused by natural disasters in the Czech Republic between 1997 and 2010. Transp. Res. Part A Policy Pract. **80**, 90–103 (2015). https://doi.org/10.1016/J.TRA.2015.07.006
41. Vink, P. (ed.): World disasters report 2013: focus on technology and the future of humanitarian action (2013)
42. Mauthe, A., Hutchison, D., Cetinkaya, E.K., Ganchev, I., Rak, J., Sterbenz, J.P.G., Gunkelk, M., Smith, P., Gomes, T.: Disaster-resilient communication networks: principles and best practices. In: 2016 8th International Workshop on Resilient Networks Design and Modeling (RNDM), pp. 1–10. IEEE (2016)
43. Ali, K., Nguyen, H.X., Quoc-Tuan Vien, Shah, P.: Disaster management communication networks: challenges and architecture design. In: 2015 IEEE International Conference on Pervasive Computing and Communication Workshops (PerCom Workshops), pp. 537–542. IEEE (2015)
44. Government of India: DISASTER MANAGEMENT Ministry of Home Affairs (2011)
45. Panteli, M., Mancarella, P.: Influence of extreme weather and climate change on the resilience of power systems: impacts and possible mitigation strategies. Electr. Power Syst. Res. **127**, 259–270 (2015). https://doi.org/10.1016/J.EPSR.2015.06.012
46. Moreno, J., Shaw, D.: Community resilience to power outages after disaster: a case study of the 2010 Chile earthquake and tsunami. Int. J. Disaster Risk Reduct. **34**, 448–458 (2019). https://doi.org/10.1016/J.IJDRR.2018.12.016
47. Basu, S., Bhattacharjee, S., Roy, S., Bandyopadhyay, S.: SAGE-PRoPHET: a security aided and group encounter based PRoPHET routing protocol for dissemination of post disaster situational data. In: ACM International Conference Proceeding Series (2015)

48. Fall, K.: A delay-tolerant network architecture for challenged internets. In: Proceedings of the 2003 Conference on Applications, Technologies, Architectures, and Protocols for Computer Communications—SIGCOMM '03, p. 27 (2003). https://doi.org/10.1145/863956.863960
49. Fall, K., Iannaccone, G., Kannan, J., Silveira, F., Taft, N.: A Disruption-Tolerant Architecture for Secure and Efficient Disaster Response Communications. https://www.semanticscholar.org/paper/A-Disruption-Tolerant-Architecture-for-Secure-and-Fall-Iannaccone/a4340ff8e0e5a9a03b06dd77f5984ea25778001a (2010)
50. Chlamtac, I., Conti, M., Liu, J.J.-N.: Mobile ad hoc networking: imperatives and challenges. Ad Hoc Netw. **1**, 13–64 (2003). https://doi.org/10.1016/S1570-8705(03)00013-1
51. Misra, S., Saha, B.K., Pal, S.: Opportunistic mobile networks: advances and applications
52. Sobin, C.C., Raychoudhury, V., Marfia, G., Singla, A.: A survey of routing and data dissemination in Delay Tolerant Networks. J. Netw. Comput. Appl. **67**, 128–146 (2016). https://doi.org/10.1016/J.JNCA.2016.01.002
53. Vasilakos, A., Zhang, Y., Spyropoulos, T.: Delay Tolerant Networks: Protocols and Applications. CRC Press (2012)
54. Pentland, A., Fletcher, R., Hasson, A.: DakNet: rethinking connectivity in developing nations. Computer (Long Beach, California) **37**, 78–83 (2004). https://doi.org/10.1109/MC.2004.1260729
55. Ntareme, H., Zennaro, M., Pehrson, B.: Delay tolerant network on smartphones. In: Proceedings of the 3rd Extreme Conference on Communication The Amazon Expedition—ExtremeCom '11, pp. 1–6. ACM Press, New York, NY, USA (2011)
56. Paul, P.S., Saha, S., Ghosh, B.C., Nandi, S., De, K., Chakraborty, S.: Demo: pSync—a peer-to-peer sync tool for challenged networks. In: 10th ACM MobiCom Workshop on Challenged Networks, CHANTS 2015, pp. 61–62 (2015). https://doi.org/10.1145/2799371.2799375
57. Paul, P.S., Ghosh, B.C., De, K., Saha, S., Nandi, S., Saha, S., Bhattacharya, I., Chakraborty, S.: On design and implementation of a scalable and reliable Sync system for delay tolerant challenged networks. In: 2016 8th International Conference on Communication Systems and Networks, COMSNETS 2016, pp. 1–8 (2016). https://doi.org/10.1109/COMSNETS.2016.7439949

Chapter 2
Post-disaster Map-Based Navigation Service

It has been observed that map-based navigation service becomes a critical prerequisite for post-disaster response operations. Such service guides volunteers to reach their destinations, helps victims to locate the shelters and supports disaster response agencies in forwarding their resources in the disaster-struck areas. However, the existing routes for reaching different destinations alter significantly in the disaster-struck areas in disaster aftermath due to the road blockage/destruction caused by disasters. Hence, the existing route maps of the disaster-struck areas become outdated. Moreover, the suitability of using digital mapping systems (web-based) remains uncertain in such areas owing to the unstable communication network in the disaster aftermath. Thus, in this chapter, Post-disaster Map Builder [1], a crowdsenced system for constructing a digital route map of the disaster-struck areas using opportunistic networks, is proposed. The proposed system constructs digital route maps of the disaster-struck areas in the mobile handheld devices (smartphones) from the GPS traces accumulated by the volunteers carrying such devices.

2.1 Literature Review

Numerous researches have been conducted in this decade for the automatic construction of route maps from GPS traces. Such researches can be mostly categorized into three, that is, point clustering [2, 3], incremental track insertion [4–6] and intersection linking [7, 8].

S. Bhattacharjee et al., *Post-disaster Navigation and Allied Services over Opportunistic Networks*, Smart Innovation, Systems and Technologies 228,
https://doi.org/10.1007/978-981-16-1240-4_2

2.1.1 Point Clustering

This method groups the GPS traces in diverse techniques to recognize the group seeds which usually represent the road junctions. The seeds are then connected to construct the portions of the route maps.

In one such work by Edelkamp et al. [2], an approach is proposed to construct high accuracy route maps with detailed information on lanes. However, this approach assumes the availability of fine-grained GPS traces as a prerequisite. This is the maiden technique to infer route maps depending on the k-means clustering algorithm.

The method illustrated in [2] is further enhanced by Schroedl et al. in [3]. In this work, a process is proposed to improve the discovery of road junctions. The improvement is attained by recognizing road junctions. Traces moving through a road junction are clustered by entry and exit points. A spline-fitting method is further employed to create the ultimate positions of the road junctions.

2.1.2 Incremental Track Insertion

This method creates route maps by progressively injecting path segments into a vacant map. Precisely, this approach uses the first path segment as a base map and improves it progressively by incorporating more segments. The map details are enhanced after each such insertion of path segments. The existing route positioning is then corrected by applying interpolation.

In one such work by Cao et al. [4], an incremental track insertion method is introduced which demonstrates the route map as a directed embedded graph. The method employs a pre-processing step that cleanses the input path segments from noise. The path segments are then injected progressively into the route map by using distance and direction as criteria. This method demonstrates an acceptable outcome only on thickly sampled trajectory traces. It does not perform satisfactorily on sparse and noisy traces.

In another work by Ahmed et al. [5], a method is introduced that implements the Fréchet distance to partially match the path segments to the map. In this method, one path segment is added to the map in three stages. The initial stage accomplishes a fractional map-matching of the path segment to the partly created map in order to recognize equal and unequal portions. In the second stage, the unequaled portions of the path segments are injected into the partly created map, by generating fresh vertices and producing and dividing edges. In the final stage, the matched portions of the path segments that cover the prevailing edges in the map are updated using a minimum-link algorithm.

Based on the effort of [4], a method for producing route maps from GPS traces is introduced by Wang et al. [6]. In this work, the authors present circular boundaries to separate points adjacent to road junctions and use a k-means clustering method to identify the junctions. This technique also assumes the availability of thickly sampled GPS traces.

2.1.3 Intersection Linking

This approach stresses on precisely locating the junction points of road segments and subsequently connecting such points. Unlike other map building methods, this approach identifies the junction points at the initial stage and the points are connected to form edges in future stages. The features such as movement speed, the direction of movement or concentration of points play a crucial role in recognizing junctions.

In one such work by Fathi et al. [7], the path segment crossings are identified from GPS traces by applying a classification algorithm. Once the crossings are identified, the raw traces are used to connect the junctions. This approach exhibits satisfactory performance in identifying simple crossings on the road networks under thickly sampled GPS traces.

In another work by Karagiorgou et al. [8], a vehicle-based route map construction algorithm is proposed. The road junctions in these algorithms are detected by identifying the changes in direction and a speed reduction to a particular limit by the vehicles. The junction points from different path segment traces are subsequently clustered to identify the road twists. Finally, the junction points are connected to construct the road segments. This technique requires relatively less densely sampled data (>30 s).

The aforesaid works described in [2–8] consider unrealistic assumptions regarding the trace data such as less noise, high sampling frequency and uniform distribution of GPS traces, which suggests data are collected by moving through a path segment exactly once. In addition to that, most of the aforesaid works assume the availability of thickly sampled GPS traces from the trace collection vehicles using survey-grade GPS devices [9]. However, in reality, the GPS traces are collected by means of walking or using non-motorized vehicles (bicycles) in the disaster aftermath [10]. In addition to that, the GPS receivers of mobile handheld devices such as smartphones, tablets, etc. are typically used for trace collection in such scenarios. Such GPS receivers lack the required accuracy in accumulating fine-grained traces compared to survey grade GPS receivers. Hence, some of the works on constructing route maps from sparsely sampled trajectory traces are reviewed below.

In another such work by Kasemsuppakorn et al. [11], pedestrian collected GPS traces are used in route map construction. The authors design an algorithm for automatically constructing route maps from multiple GPS traces. The GPS traces accumulated in this work are mostly free from noise as the pedestrian volunteers move continuously while collecting traces. However, this notion of uninterrupted movement turns out to be unrealistic in most real-life scenarios such as post-disaster scenarios.

In order to overcome the restrictions of the earlier work, Blanke et al. [12] suggest a method for creating route maps from the crowdsourced GPS traces. This approach deals with noisy GPS traces generated due to the high disparity in movement activities by the volunteers. However, this method involves a large-scale dataset to construct route maps which might be unavailable in a post-disaster scenario.

In another work by Qiu et al. [13], a method is proposed to build route maps from the thinly sampled GPS traces by applying a segmentation-and-grouping mechanism. The primary notion of their work is to construct the path segments from traces and subsequently produce the map of the route by clustering the segments. Their proposed algorithm deals with noisy and thinly sampled GPS data.

All the aforementioned works [2–8, 11–13] require steady network connectivity to construct route maps from GPS traces. As per our understanding, there is only one work by Trono et al. [14] which constructs route maps from GPS traces under sporadic network connectivity.

In [14], a method is proposed to build the route map of the disaster-struck areas over the opportunistic networks. The work presumes the availability of a couple of node categories such as sensor and computing nodes. The sensor nodes are primarily the mobile devices engaged to accumulate GPS traces. The traces collected by mobile nodes are propagated to the computing nodes (PCs), where the actual route maps are created. The authors employ the identical approach mentioned in [12] to build maps. In addition to that, a load balancing heuristic is also introduced for efficient distribution of the map construction process among different computing nodes. The route maps constructed by the computing nodes are then forwarded to the sensor nodes over opportunistic networks.

The works described in [2–8, 11–13] require steady network connectivity and high end computing resources to build route maps which is unrealistic in disaster aftermath. Although the sporadic network is taken into account in [14], the assumptions concerning the requirement of high end computing resources remain intact. The accessibility to such resources proves infeasible in the disaster-struck areas of developing countries like India due to the unsteady power supply and an inadequate number of trained volunteers to operate such devices. Thus, all the aforesaid map building methods [2–8, 11–14] remain inappropriate to build route maps from GPS traces in a post-disaster scenario.

2.2 System Architecture

In this section, the system architecture of the Post-disaster Map Builder including its associated assumptions is presented. There are primarily three types of nodes, namely *volunteer nodes*, *camp nodes* and *ferry nodes* in this architecture. A disaster typically does not strike a region equally. According to disaster management terminology, the severely affected portions of a region are considered as disaster-struck areas. Due to disaster, some of the roads/pathways in these affected areas may become unreachable, and therefore, it is imperative to find out other substitute routes between the shelters and the camps. Usually, the volunteers find out such routes or pathways by getting assistance from neighboring residents. The *volunteers (nodes)* collect trajectory traces of such routes with the help of GPS of their smartphones and share the traces with their fellow volunteers working in the same disaster-struck area. Here, to communicate among the nodes, the Trace Route routing protocol is proposed which is described in Chap. 5 (Sect. 5.4).

The system architecture of the Map Builder is presented in Fig. 2.1. The figure shows two disaster-struck areas in the region. Both the areas have one relief camp, one medical camp and a few shelters. There is one command and control center located at a distant place from the disaster-struck areas. In those two disaster-struck areas, ten *volunteers* are working from different camps. This figure also shows that the connecting road between medical camp M2 and one shelter becomes unreachable. Thus, volunteer V_{1M2} takes different routes for reaching the M2. It is also shown that a *ferry node* (e.g. supply vehicle) from the command and control center travels through the unaffected roads to visit the camps across the disaster-struck areas.

The *volunteer nodes* accumulate trajectory traces comprising a group of trajectory points and store the traces in their handheld smartphones. In the disaster aftermath, the *volunteer nodes*, *camp nodes* and *ferry nodes* of a disaster-struck area form a potential group amongst which such traces are exchanged. The *ferry nodes* transmit trajectory traces over long range in opportunistic networks. Whenever one such node comes in contact with another node, they share their trajectory traces with each other. The *volunteer nodes* and *ferry nodes* follow the post-disaster movement (PDM) model

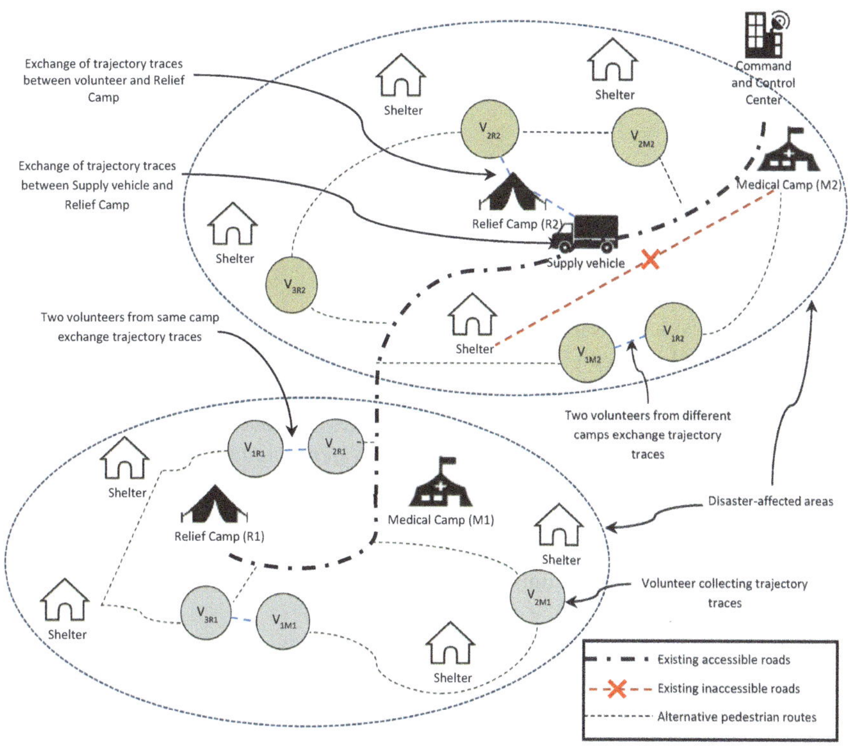

Fig. 2.1 System architecture of Post-disaster Map Builder

[15] which mimics the typical mobility characteristics of volunteers and victims in post-disaster scenarios.

In this system architecture, there are few assumptions. The *volunteer* and *ferry nodes* are considered to have GPS receivers. Each of the nodes has the same initial energy and consumes the same amount of energy for transmitting and receiving the trajectory traces. Once in a day, the *ferry nodes* visit *camp nodes* from the command and control center. To save the battery power, instead of collecting all the trajectory points, the *volunteer nodes* gather the points at a predefined threshold gap (e.g. Euclidean distance 20 m). The nodes share the traces as per the following.

- A *volunteer node* of a disaster-struck area is allowed to share traces with the *volunteer* and *camp node* located within the same area. The node is also permitted to send the traces to the *ferry nodes* and *camp nodes* across the disaster-struck areas.
- A *camp node* located in a disaster-struck area is permitted to share the traces of the area with *volunteer nodes* of the same area. However, the node is allowed to exchange the traces with *ferry nodes* across the disaster-struck areas.
- A *ferry node*, on the other hand, is only permitted to share the traces with other *ferry nodes* and *camp nodes* across the disaster-struck areas.

Following the above sharing procedure, over time, not only the *volunteer nodes* can collect the said traces from the disaster-struck areas locally but also the traces of the whole region having several disaster-struck areas are collected through the camp nodes and command/control center.

2.3 Post-disaster Map Builder Communication Architecture

This section illustrates the functionalities of different units of the Post-Disaster Map Builder system along with their interoperability. The system is mainly composed of *trace management* and *map inference* units. The communication architecture of the suggested system is illustrated in Fig. 2.2. It is assumed that the system should be pre-installed in *volunteer nodes*, *camp nodes*, *ferry nodes* and in the command and control center. The periodic accumulation of GPS traces by the *volunteer nodes* is handled by the *trace management* unit. In addition to that, the interchange of traces amid different nodes (*volunteer nodes*, *camp nodes* and *ferry nodes*) is also controlled by this unit. This unit comprises two sub-units such as *trace collection* and *trace interchange*.

The accountability for creating route maps from the accumulated GPS traces is delegated to the *map inference* unit. This unit involves three sub-units such as *pre-processing*, *significant point clustering* and *topology inference*.

Production of the complete route map of a whole region in an incremental approach includes the following steps:

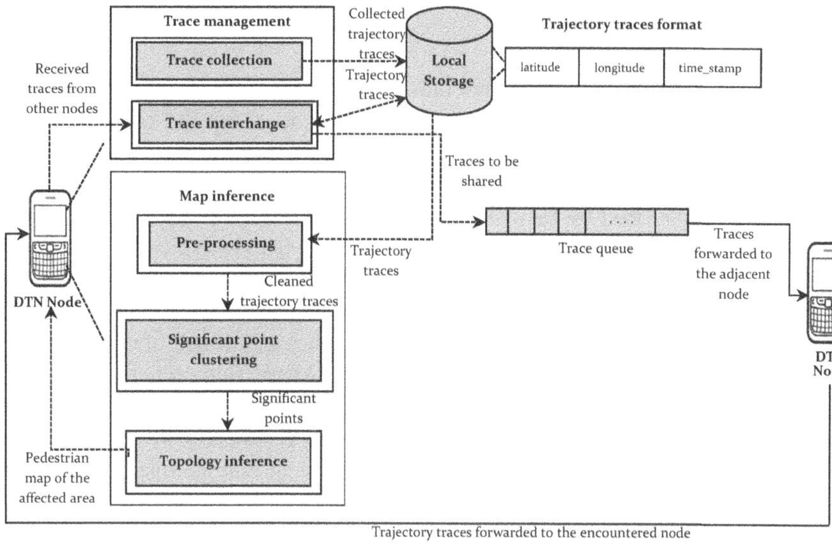

Fig. 2.2 Communication architecture of the proposed Post-disaster Map Builder system

- Building local route maps of the disaster-struck areas.
- Constructing a global route map by combining the local maps.

2.3.1 Trace Management

In order to create the local route map of a specific disaster-struck area, the GPS traces gathered by each *volunteer nodes* need to be exchanged with other identical nodes and *camp nodes* associated with that area. Such an exchange is accomplished by implementing a group-based publish-subscribe mechanism [16]. The potential publisher of GPS traces is *volunteer nodes* in typical post-disaster scenarios. On the contrary, other *volunteer* or *camp nodes* residing in a similar zone are the potential subscribers of the traces. In this fashion, the gradual collection of GPS traces from a certain area takes place at *volunteer* and *camp nodes* of that part.

Building the global route map of the whole region requires the interchange (publish-subscribe) of the GPS traces stored at *camp nodes* with *ferry nodes* and vice versa across the disaster-struck areas. The GPS traces are progressively gathered at *camp nodes* from different disaster-struck areas in this fashion. The GPS traces accumulated by the *ferry nodes* across the disaster-struck areas reach the command and control center once they return to their origin. Interchange of trajectory traces among different nodes at the cost of low network overhead necessitates group-based multicasting (*groupcasting*) of such traces. Hence, the GPS traces are forwarded using *groupcast* mode in this unit.

2.3.1.1 Trace Collection

The GPS traces consist of tracepoints indicating latitude and longitude. The periodic gathering of GPS tracepoints from *volunteer nodes* and the construction of trajectory traces from such points is taken care of by this sub-unit. The tracepoints are collected based on the shift of the location of *volunteer nodes* rather than time. The tracepoints gathered by the *volunteer nodes* are usually thinly sampled. It has been observed that, during their journey, the *volunteer nodes* typically move at an uneven speed with occasional stoppages which makes the tracepoints susceptible to random noise. In order to suppress the generation of random noise, the tracepoints are collected from the *volunteer nodes* only after the occurrence of a pre-defined displacement (20 m) from their previous positions. Apart from reducing the likelihood of random noise generation, this technique also cuts down energy consumption by eliminating the need for polling of GPS receivers periodically. Moreover, owing to the missing/bad GPS signal, occasionally it is evident that the distance between two successive tracepoints becomes much greater than the pre-defined displacement which may yield inaccurate trajectory traces. In order to suppress the gaps between trajectory traces, trajectory up to that gap is segmented and new trajectory traces are generated. Inspired by [12], the thresholding distance of 30 m is imposed among successive tracepoints for trajectory segmentation.

Traces generated by a node till a specific time can be distinctively identified by $Trace\_t_{creator\_id}$ where *creator_id* acts as a distinctive node identifier of the originator node and t is the timestamp of trajectory traces creation. Each $Trace\_t_{creator\_id}$ consists of a sequence of tracepoints (p). Each of such points is characterized by 3-tuple p = (lat, long, t) where lat and long stand for the latitude and longitude of the point acquired at timestamp t.

The GPS traces acquired by *volunteer nodes* through its GPS receiver or trace exchange with other nodes are stored in the local storages (secondary storage of mobile handheld devices) of the originator nodes. A trace queue is maintained at each node for storing the traces to be forwarded to the encountered node.

2.3.1.2 Trace Interchange

The primary responsibility of this sub-unit is interchanging GPS traces among different nodes. The exchange of traces in *groupcast* communication is identical to the epidemic forwarding of traces, as all types of nodes present in a particular disaster-struck area are the prospective receiver of such traces. However, in opportunistic networks, the epidemic forwarding protocol is intended only for unicast communication. Implementing *groupcasting* through unicast communication causes excessive network overhead and huge energy consumption. Thus, traditional Epidemic routing protocol does not yield the desired outcome in post-disaster scenarios. Accordingly, a new protocol (Trace Route routing protocol) has been proposed for exchanging GPS traces. The illustration of the Trace Route protocol is presented in Chap. 5 (Sect. 5.4). As opposed to Epidemic protocol, the Trace Route protocol forwards the trajectory traces produced by a node (*Trace*_$t_{creator_id}$) to a pre-defined group of destinations (*destination group*) depending on the position of the originator node. This protocol eradicates the likelihood of exorbitant replication of traces by dispatching them toward their prospective recipients. Thus, the communication overhead is cut down significantly by employing the aforesaid protocol causing a substantial reduction in energy requirements of the entire network. The algorithm of the *trace management* unit is as follows.

Algorithm: Trace management
Input: Trajectory traces (*Trace*_$t_{creator_id}$) and Trace list of the originator node (*Trace*_$list_{source}$)
Output: Exchange of trajectory traces (*Trace*_$t_{creator_id}$) between encountered nodes
Algorithm Triggered: Upon request from the user

```
Begin
   call trace_collection(p_last)                    /* collect trajectory traces based on displacement */
   call trace_interchange(node_id_encountered)

/*Function*/
trace_collection(p_last)
displacement_limit = 20                             /* node displacement_limit is set to 20 meters */
segmentation_threshold =30                          /*segmentation_threshold is set to 30 meters */
d = Euclidean_Dist(p_now,p_last)                    /*d=Euclidean distance between p_now and p_last */
if (d≤segmentation_threshold) then
   if (d≥displacement_limit) then
      temp ←p_now                                   /*store current trajectory point (p_now) in temp*/
      p_last = p_now
   end if
end if
Trace_t_creator_id←temp                             /*store temp in Trace_t_creator_id*/

/*Function*/
trace_interchange(node_id_encountered)
fetch Trace_list_encountered                        /* by exchanging trace list with encountered node */
trc_list[i ] = set_difference(Trace_list_source,Trace_list_encountered) */ list of trajectory traces */
for eachTrace_t_creator_id in trc_list[i]
   fetch dest_group[j ]                             /* destination group of trajectory traces Trace_t_creator_id */
   while(dest_group[j ]is not empty) do
      if (node_id_encountered ==dest_group[j] then  /* encountered node is in the destination list of T_id */
        queue ←Trace_t_creator_id                   /* insert trajectory trace into trace queue */
      end if
      j++                                           /* incrementing loop counter */
   end while
end for
while (queue is not empty) do
  forward trajectory traces to node_id_encountered
end while
End
```

The Trajectory traces (*Trace*_$t_{creator_id}$) and Trace list (*Trace*_$list_{source}$) of the source node to be shared are the inputs to the algorithm. Identifying suitable encountered nodes for exchanging of trajectory traces (*Trace*_$t_{creator_id}$) is the outputs from the algorithm.

2.3.2 *Map Inference*

This unit is installed in every node in the network and is responsible for building route maps from the GPS traces. The unit is designed in such a way to make the map building procedure computationally less demanding and energy efficient as the route maps are created with low-powered mobile devices. The entire map construction task is performed by three sub-units presented as follows.

2.3.2.1 Pre-processing

The primary function of this sub-unit is cleansing and leveling GPS traces from different noises. We design a low resource and low energy-hungry data-driven algorithm for leveling the traces. The precision of mobile devices accumulated GPS traces in open-air should vary between 5 and 8.5 m [17]. Though, in reality, occasionally this variation may reach up to 30 m due to the occurrence of errors from multiple sources. The different causes of errors in the GPS traces are elaborated below.

- **Warm start/cold start problem**: In order to compute positions precisely, the GPS receivers usually necessitate signals from at least four satellites. This signal attainment procedure is time-consuming causing occasionally absent points at the start of the GPS traces.
- **Satellite clocks**: The imprecision in satellite timers sometimes causes decreased precision in setting the GPS points.
- **GPS receiver clock**: The imprecision in the receiver's timer decreases the precision of GPS points.
- **Atmospheric effects**: Turbulence in the ionosphere and variations in the atmosphere disturb the speed and route of GPS signals instigating imprecisions in locating the GPS receiver.
- **Multipath effects**: The GPS signal is habitually redirected by skyscrapers, walls or surfaces in metropolitan regions. Thus, the predicted locations of the GPS receivers are typically dislocated from their actual positions.
- **Signal blocking**: GPS signals can be obstructed easily by ingredients like plastic, metal, etc. due to their frail signal strength. Such obstructions produce lost/misplaced GPS points.

Although some mechanisms for error reduction caused by the satellite timers and GPS receiver timers are incorporated in the regular GPS receivers, the elimination of errors exclusively from GPS traces is practically unachievable in reality. Therefore, the traces acquired from the GPS receivers need cleansing prior to map creation. In practice, regular leveling methods, such as kernel smoothers, smoothing splines, generalized additive models and Kalman filters, are extensively employed to reduce errors from the traces. However, these approaches are usually prediction-based which may alter the exact orientation of the GPS traces. Therefore, a data-driven (statistical) approach for leveling traces is of utmost importance.

Chasel et al. [18] designed a data-driven trajectory leveling method that levels GPS traces by inserting each point p of the trace into a high-dimensional space as p' utilizing model points before and after p. This procedure of converting p into p' is termed as the lifting of p. Each point p' in the elevated copy of the trajectory is then steered to its adjacent neighbors in the elevated space (Laplacian smoothing). Finally, the actual traces are restored by calculating the arithmetic mean of the appropriate coordinates of the elevated points. But, leveling each point in the GPS traces is infeasible as it requires an enormous amount of computing power and energy consumption. In addition to that, numerous trajectory traces of the same path segment is required for accurate leveling of GPS traces which in reality is an impractical assumption in disaster aftermath. If multiple GPS traces of a matching path segment is unavailable, this method merely computes a moving average of adjacent points to attain an even trajectory. The method of calculating the moving average modifies the genuine positioning of the underlying path by eradicating topographic particulars from the tracepoints. Hence, this technique needs to be modified to preserve the topography of GPS traces. We adapt the leveling technique proposed by Chasel et al. [18] by implementing a selective lifting of points from GPS traces in order to diminish the effect of computation overhead, energy consumption and retain the real topology of the underlying path.

Our suggested trajectory leveling (pre-processing) mechanism primarily detects the GPS points having noise in $Trace\_t_{creator\_id}$. The presence of noise in a tracepoint is detected by measuring the bearing change ($\Delta\alpha$) between successive tracepoints.

The GPS receivers installed in mobile devices typically offer bearing information. We do not consider such bearing information as it does not offer the required precision, particularly while walking (movement speed of the nodes is less than 3.0 m/s) is the principal means of trace accumulation [11]. On the contrary, we compute the bearing change ($\Delta\alpha$), by subtracting consecutive bearings using the great circle navigation formula based on Eqs. (2.1) and (2.2).

Thus, bearing between two consecutive points α can be calculated as follows:

$$\alpha = \tan^{-1} 2(\sin\ \Delta\text{long} \cdot \cos\ \text{lat}_2, \cos\ \text{lat}_1 \cdot \sin\ \text{lat}_2 - \sin\ \text{lat}_1 \cdot \cos\ \text{lat}_2 \cdot \cos\ \Delta\text{long}) \quad (2.1)$$

where $(\text{lat}_1, \text{long}_1)$ and $(\text{lat}_2, \text{long}_2)$ are the latitude and longitude of the points, $\Delta\text{long} = (\text{long}_1 - \text{long}_2)$ is the variation between longitudes of two points. The bearing change $\Delta\alpha$ is calculated as:

$$\Delta\alpha_{12} = |\alpha_1 - \alpha_2| \quad (2.2)$$

where α_1 and α_2 are bearings. As per our assumption, the tracepoints are exposed to noise if repeated bearing changes beyond a selected threshold (30°) exists between the consecutive points of GPS trace.

The value for bearing change usually varies between 0° and 360°. A 12-direction chain code, as illustrated by Kasemsuppakorn et al. [11], has been employed in the suggested pre-processing technique to signify the sharpness of the road curves.

Fig. 2.3 Pictorial representation of 12-direction chain code

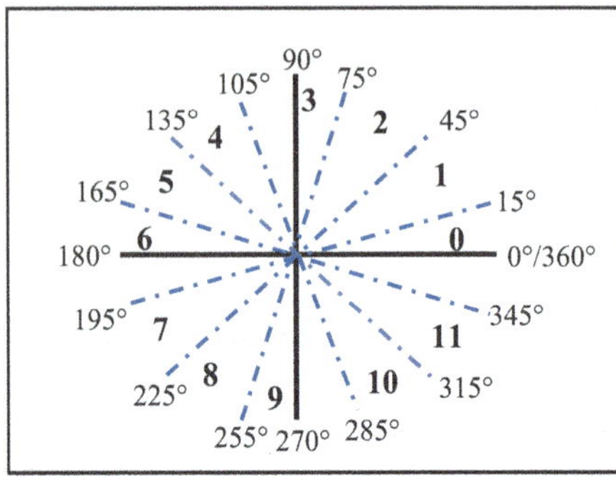

Apart from that, the 12-direction chain code is capable of detecting the deviation in path directions precisely. A pictorial representation of the 12-direction chain code is illustrated in Fig. 2.3.

A noisy point can be recognized by our suggested pre-processing technique as follows:

- If more than solitary bearing change beyond a pre-defined threshold (30°) is present around four adjacent trajectory points in $Trace\_t_{creator\_id}$, then the third point is vulnerable to noise.
- Such noisy points are only elevated to the higher dimensional space with their previous and succeeding points, respectively.
- The noise is eradicated by calculating the arithmetic mean on the relevant coordinates of the elevated points.

Thus, the substantial reduction in computation overhead and energy requirement can be achieved at nodes by detecting and elevating merely the noisy points to the higher dimensional space.

An explanatory example of the suggested pre-processing technique is depicted in Fig. 2.4. In this example (Fig. 2.4a) the path segments between points P_1 and P_2 and

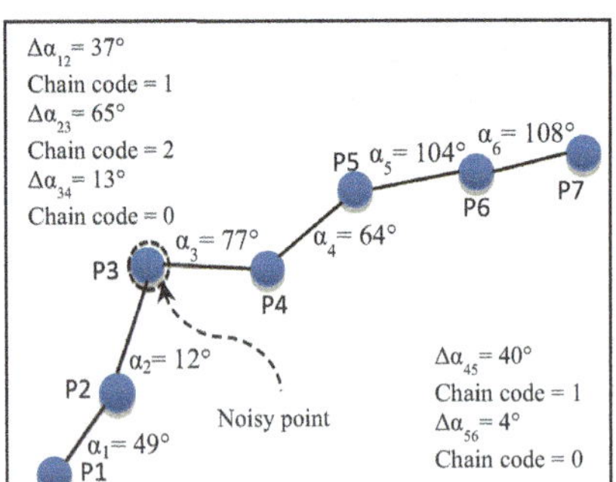

(a) An instance of identifying the noisy point

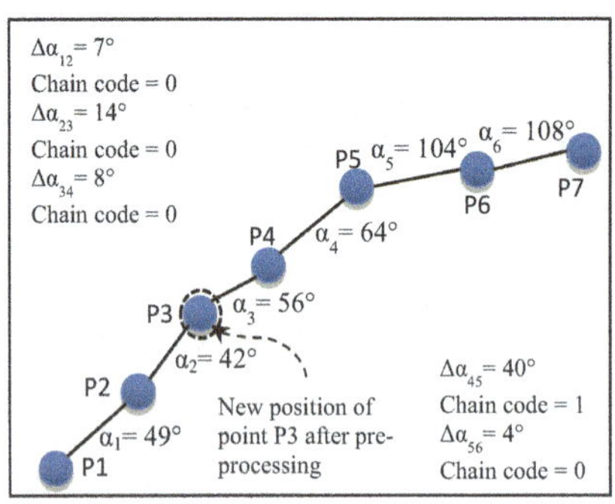

(b) Sample trajectory trace after smoothing

Fig. 2.4 An illustration of the proposed pre-processing technique between P_2 and P_3 have bearing values α_1 (49°) and α_2 (12°), respectively. Hence, the bearing change $\Delta\alpha_{12}$ around the point P_2 is 37° which resembles chain code 1. Also, the $\Delta\alpha_{23}$ around point P_3 is 65° which resembles chain code 2.

Thus, a couple of successive bearing changes (beyond 30°) of 37° and 65° occurs about successive points P_1, P_2, P_3 and P_4. Thus, it has been identified that the point P_3 contains noise. In addition to that, a bearing change ($\Delta\alpha_{45}$) of 40° occurs around point P_5. This point is not identified as noisy since there is no successive bearing change around that point. Figure 2.4b portrays an instance of GPS trace points after noise elimination. The variation in the location of point P_3 clearly illustrates the efficiency of the pre-processing mechanism. This sub-unit is automatically initiated after initiating the map inference unit. The stored GPS traces in a node are treated as inputs to this unit and the cleansed traces are treated as outputs.

2.3.2.2 Significant Point Clustering

This clustering works in two steps. First, it identifies some points from every trajectory traces, which hold important data regarding the path segment. These identified points are considered significant points as these are crucial in deciding the topology of a path segment from the trajectory traces.

In the second step of such clustering, it clubs the significant points, identified in the first step, of multiple trajectory traces of a specific path segment to create representative significant points. Such multiple trajectory traces may be collected in two different ways—traversal of *volunteer nodes* through the same path segment and traversal of one *volunteer node* through the matching path segment in several instances. This clustering of significant points removes imprecision which may be present even after doing pre-processing of the trajectory traces.

Most of the current literature on map inference works on gathering traces of moving transports. Here, discovering road twists from the collected traces is not very difficult as the vehicles typically change their velocities and directions in road twists. However, in our application scenario, *volunteer nodes* typically collect the traces at walking speed. This makes discovering road twists from the changes in movement speeds unfeasible. So, we propose a method to identify significant points from the traces based on bearing change ($\Delta\alpha$). We use the 12-direction chain code [11] to fix a threshold value of bearing change. The $\Delta\alpha$ is computed using Eqs. (2.1) and (2.2). The proposed method finds significant points as below:

- In the trace, if the $\Delta\alpha$ value of a point is found beyond a threshold, say 30°, it implies a road twist and the point is assumed as a significant point.
- The initial and final points of each of the traces are assumed as significant points.

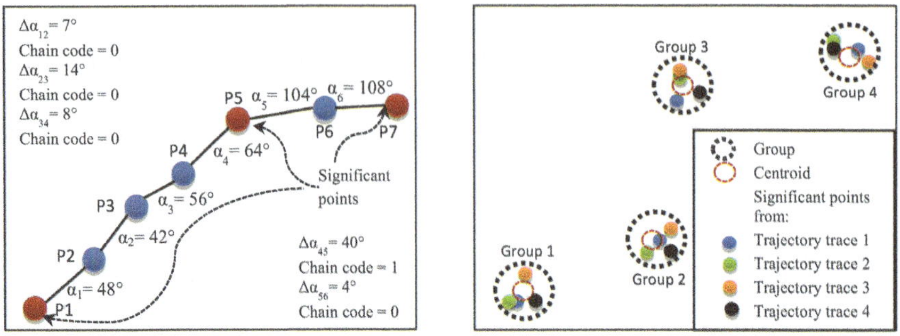

(a) Identification of significant point from a trajectory trace

(b) Clustering significant points from multiple trajectory traces

Fig. 2.5 An illustration of significant point clustering

Figure 2.5a illustrates an example of significant point identification. In this example, the path segments between points P_4 and P_5 and between P_5 and P_6 have bearing values α_4 (64°) and α_5 (104°), respectively. Hence, the bearing change $\Delta\alpha_{45}$ around the point P_5 is 40° which corresponds to chain code 1. Thus, P_5 signifies a road twist implying a significant point.

In the second step of clustering, we do not adopt any conventional clustering techniques which are computationally intensive. Instead, it is done by computing the centroid of each of the matching groups of significant points. To do this, at first, the significant points of a specific trajectory trace are selected as preliminary centroids of each group. The significant points of the other traces from the matching path segments are included one by one into a group based on the criteria of group selection. A significant point is considered to be included in a group if the Euclidean distance between the point and the initial centroid of the group is within 10 m. After such inclusion of each one, the centroid of the group is recomputed. Once the inclusion of all the significant points from all the traces of a matching path is completed, the centroids of the groups are considered as the representative significant points of all the traces. Such grouping/clustering of significant points from four traces is illustrated in Fig. 2.5b.

2.3.2.3 Topology Inference

This unit takes representative significant points as input and generates the route map of the disaster-struck areas by joining such points. Two adjacent significant points can be connected subject to the fulfillment of the following:

- If two adjacent significant points belong to a matching path segment and either of them is an initial or final point.
- If $\Delta\alpha$ around the points is above 30° (threshold value) and neither of them is an initial or final point of the path segment.

In one example, the topology inference of a path segment with no obstruction is shown in Fig. 2.6a. Here, two adjacent points P_1 and P_2 are connected as both of them correspond to the same path, and P_1 is the initial point of the path. Similarly, the adjacent points P_2 and P_3 are connected as P_3 is the final point of the path.

In the other example, the topology inference of a path segment with obstruction is shown in Fig. 2.6b. Here, two adjacent points P_1 and P_2 are connected as both of them correspond to the same path, and P_1 is the initial point of the path. But, although P_2 and P_5 are adjacent points, these are not connected. The reason is the value of $\Delta\alpha_{15} = 6°$ around P_2 and therefore, it does not indicate any road twist.

Once the significant point clustering produces representative significant points, the topology inference unit is activated. The map inference algorithm is given below.

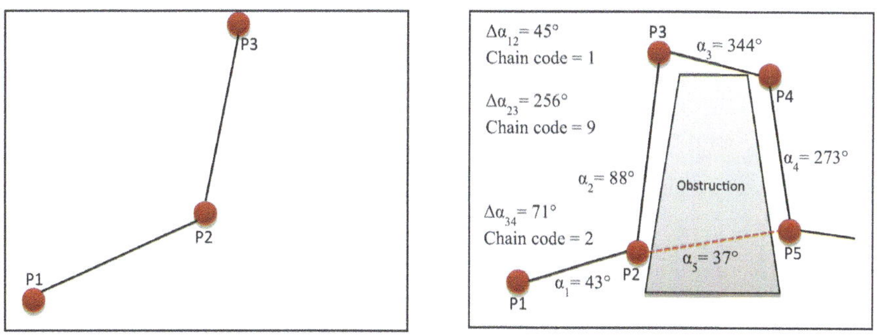

(a) An instance of the path without obstruction

(b) An instance of the path with an obstruction

Fig. 2.6 An illustration of topology inference

Algorithm: Map inference
Input: Trajectory traces ($Trace\_t_{creator\_id}$) stored in a node
Output: Route map
Algorithm Triggered: Upon request from the user

```
Begin
for each Trace_t_creator_id in Trace_list_node_id
   call preprocessing(Trace_t_creator_id)
   call significant_point_clustering(Trace_t_creator_id)
end for
   call topology_inference(sgp_table)

/*Function*/
preprocessing(Trace_t_creator_id)
for each p in Trace_t_creator_id
   Compute Δα                          /*compute bearing change between consecutive points based on equation (2.2) */
   if (p containing noise)then
      lift p to p'                     /*lifting p to a higher dimensional space */
      p = arithmatic_mean(p')          /* perform arithmetic mean on p' and assign it to p */
      Trace_t_creator_id ← p           /* store p in the trajectory trace */
   end if
end for

/* Function*/
significant_point_clustering(Trace_t_creator_id)
for each p in Trace_t_creator_id
   Compute Δα                          /* Compute bearing change between consecutive points based on equation (2.2) */
   Compute C_c                         /* Compute the chain code value */
   if (C_c > 0 || 6) then              /* identify significant point as per 12-direction chain code */
      sgp_table ← p                    /* store p in significant points table (sgp_table) */
   end if
end for
compute centroid from identical groups of significant points
/* Function*/
topology_inference(centroids[k])
for each p in centroids[k]             /* adjacent neighbor of p from similar path segment */
   find p_near
   if (p or p_near is a starting point or endpoint) then
      connect p and p_near
   end if
   if (p or p_near is not a starting point or endpoint) then
      compute Δα
      compute C_c
      if (C_c > 0 || 6 around both p and p_near) then
         connect p and p_near
      end if
   end if
end for
End
```

2.4 Performance Evaluation

As the performance of the *trace management* unit is dependent (Sect. 2.3.1.2) on its underlying protocol Trace Route, the performance evaluation of this unit will be illustrated in Chap. 5. Hence, the performance of the *map inference* unit is evaluated in this section with the help of simulation. The simulation of this is implemented in the Mapping Toolbox of MATLAB 2017a.

2.4.1 Simulation Setup in MATLAB

The simulation environment for assessing the performance of the *map inference* unit is set up in the Mapping Toolbox of MATLAB 2017a which contains essential functions and algorithms for analyzing GPS traces and produce route maps from such traces. Initially, the GPS traces accumulated by a volunteer are imported to the toolbox by employing *gpxread()* function. Our suggested *map inference* algorithm is executed in the later stage. Lastly, the route map is created and overlaid on the Open Street Map of the disaster-struck areas with the help of *geoshow()* and *webmap()* functions of the toolbox.

2.4.2 Simulation Metrics

The *map inference* unit, whose predominant purpose is to create route maps from the GPS traces, is evaluated by analyzing *root mean square* (RMS) error. It is a characteristic dimension [19] to estimate the precision of the constructed route maps. However, the characteristic estimation of the constructed route maps is not absolutely appropriate owing to the absence of relief maps of the disaster-struck areas as reference data. The obtainable relief maps of such areas turn out to be outdated due to the alteration of existing road networks caused by disasters. Hence, such maps could not be treated as a reference while estimating the perfection of the constructed route maps. However, for the extensiveness of the measuring accuracy of the route maps the offline Open Street Maps of the disaster-struck areas are considered as a reference. The RMS error specifies the mean deviation amid the constructed path segments and the actual path segments.

Root mean square error: The arithmetic mean of displacement amid the constructed path segments and the actual path segments are specified by RMS error. The lower RMSE value implies improved trajectory precision of the constructed path segment. The metric is defined as follows:

$$\text{RMS error} = \sqrt{\frac{\sum_{i=1}^{l} d(\text{cnst}_i, \text{act}_i)^2}{l}}$$

Here, l is the number of path segments; $d(\text{cnst}_i, \text{act}_i)$ is the distance between constructed path segment and the actual path segment.

2.4.3 *Results and Discussion*

This section analyzes the simulation results of the Post-disaster Map Builder system under diverse scenarios. There are two important units in such a builder: *trace management* and *map inference*. The performance of the *trace management* unit is dependent (Sect. 2.3.1.2) on its underlying protocol Trace Route which will be presented in Chap. 5. Thus, in this section, only the performance of the *map inference* unit is evaluated.

2.4.3.1 Viability and Efficacy Evaluation of *Map Inference* Unit

Here, we evaluate to what extent our proposed *map inference* unit is viable in making trajectory trace-based digital route maps. Let us analyze the viability of the route map generation from collected traces as shown in Fig. 2.7. We consider four sample raw trajectory traces containing 230 trajectory points. The raw trajectory traces and the route map generation from these traces are shown in Fig. 2.7a, b, respectively. We see from Fig. 2.7b that there are noises in the inferred paths. Now, *pre-processing* is done on the inferred paths to get the traces as in Fig. 2.7c. Here, we observe the equal number of points as in raw trajectory traces. Now, based on the *pre-processed* traces the route map is inferred as in Fig. 2.7d. In this figure, we see that the path segments still have several noisy points.

Subsequently, *significant point clustering* is applied to *pre-processed traces* to get important sample points as shown in Fig. 2.7e. The sample points are approximately 13% of the initially gathered traces, which is a significant reduction in the number of tracepoints. This reduces computation overhead in the inferring path from these points, thereby requiring lesser energy to get the inferred path which is shown in Fig. 2.7f. We observe from the figure that the path does not have any superfluous points. Thus, the proposed *map inference* unit is able to make route maps of the disaster-struck areas from the said tracepoints.

We measure the efficacy of the proposed *map inference* unit by comparing it with an existing map inference scheme by Karagiorgou et al. [8]. This competitor scheme uses intensive computing resources for implementing their inference scheme. We consider RMS error (RMSE) as the measuring metric which basically measures the spatial gap between the generated pathway and the existing pathway. The accuracy in terms of RMSE of both the competitor scheme is plotted with varying numbers of repeated trajectory traces in Fig. 2.8. We observe from the figure that the RMSE decreases with an increasing number of repeated trajectory traces for both the schemes. This implies that accuracy increases with the number of repeated trajectory traces. We also notice that both schemes give more or less the same accuracy to generate path segments.

The other interesting observation is that the accuracy considerably improves up to five number of repeated trajectory traces. But beyond this point, it doesn't improve much. This observation helps us in fixing a guideline of using at least five trajectory

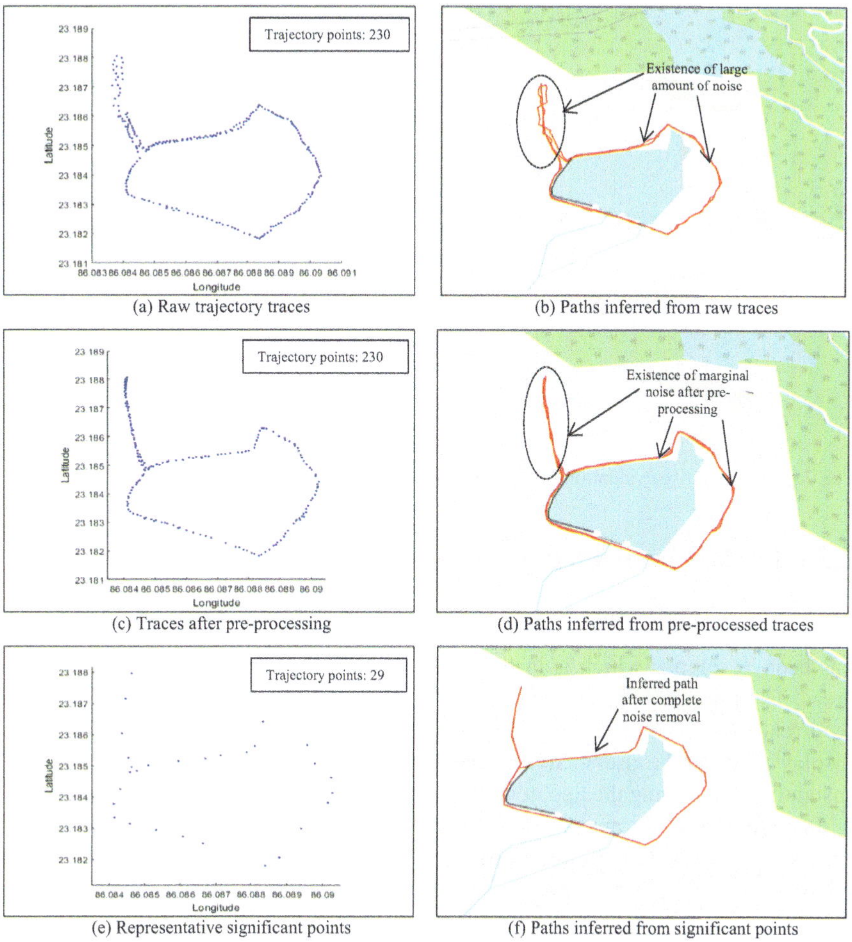

(a) Raw trajectory traces
(b) Paths inferred from raw traces
(c) Traces after pre-processing
(d) Paths inferred from pre-processed traces
(e) Representative significant points
(f) Paths inferred from significant points

Fig. 2.7 An illustration of the route map constructed by the *map inference* unit

traces for each path segment toward making a high accuracy route map from trajectory traces.

Summarily, we claim that the proposed Map Builder system is viable in making route maps even in the absence of a high-end computing device. We further claim that our proposed system attain accuracy near-identical to a system which works in a conventional scenario in the aftermath of a disaster. This inspires us to develop an Android application (app) based on the Post-disaster Map Builder system.

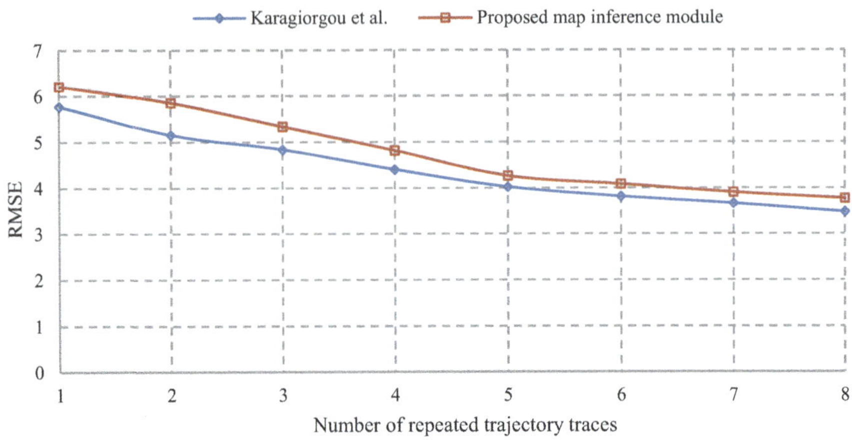

Fig. 2.8 Comparative evaluation of the *map inference* unit

2.5 Testbed Implementation

Based on the communication architecture described in Sect. 2.3, an Android app has been developed which is installed at all nodes that exist in the disaster-struck areas. The primary objective of the app is to acquiring and distributing GPS traces recurrently. Moreover, the app can produce route maps from the GPS traces and display them to the users. Some snapshots of the app are illustrated in Fig. 2.9. Instantly after starting the app, few buttons such as SHOW MAP, START SERVICE and STOP SERVICE appear on the home page, as shown in Fig. 2.9a. The GPS trace collection and distribution functions can be initiated by pressing the START

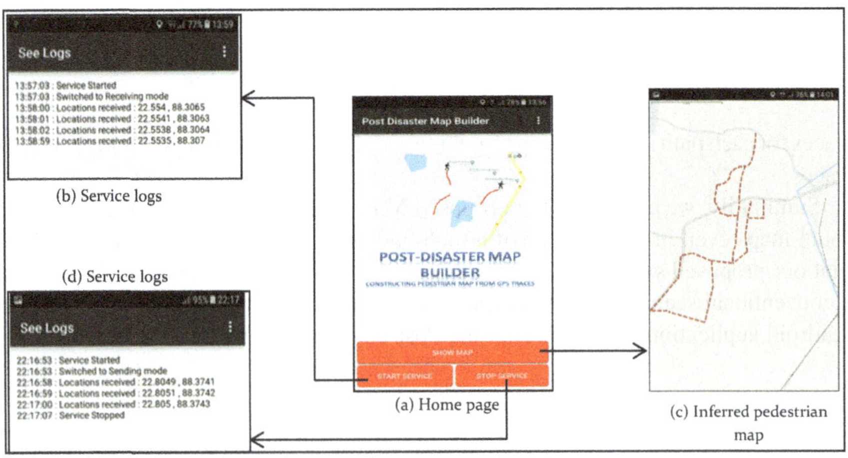

Fig. 2.9 Snapshots of the Post-disaster Map Builder app

SERVICE button. This functionality keeps on executing in the background after initiation and can be terminated only by pressing the STOP SERVICE button.

Figure 2.9b specifies the sample record of GPS traces collected by the app over time. The .gpx files are typically used for preserving the GPS traces. The constructed route map can be made visible to the user by pressing the SHOW MAP button. An instance of a route map (drawn using red dotted lines) created by the app is pictorially depicted in Fig. 2.9c. The constructed route map is overlaid onto a pre-cached map. In this app, the Open Street Map (OSM) in offline mode is employed for obtaining the pre-cached map image. The *osmdroid* Android class is engaged in integrating the offline OSM with the app.

The background execution of the app can be terminated by applying the STOP SERVICE button on the home page. Figure 2.9d demonstrates the sample status record of the app after applying the STOP SERVICE button.

2.5.1 Testbed Setup

We organize numerous field trials at Purulia and Bankura districts, West Bengal, India to assess the suitability and competence of the Post-disaster Map Builder app under diverse situations. Eleven volunteers with Post-disaster Map Builder app pre-installed in smartphones or tablets participate in the field trials. Two volunteer groups comprising five volunteers in each group are formed and positioned in such a way that direct intergroups communication does not take place. It is assumed that every group is associated with a disaster-struck area. One volunteer from every group having a tablet remains stationary and treated as *camp nodes*. One volunteer equipped with a smartphone visits the stationary nodes repeatedly on his bicycle to accumulate intergroup GPS traces. Such volunteers are treated as *ferry nodes*. The *ferry node* preserves the GPS traces in the auxiliary storage of the smartphones used by them. Interchange of GPS traces between the *ferry node* and *camp node* after the *ferry node* comes at a new group. An illustrative positioning of two volunteer groups captured during the field trial conducted at Purulia, West Bengal, India is depicted in Fig. 2.10.

During the field trials, the mobile devices manufactured by diverse vendors are used to evaluate the cross-platform connectivity of the app. The degree of distribution of GPS traces over time and energy consumptions are considered as two key parameters for gauging the efficiency of the app.

The energy intake of the Post-disaster Map Builder app is detected by an Android app called Trepn profiler [20]. The energy intakes from the Trepn profiler have been logged during the interchange of GPS traces by the Post-disaster Map Builder. The arithmetic mean of the energy requirements from all the devices employed in the field trial is considered as the average energy consumption by the app.

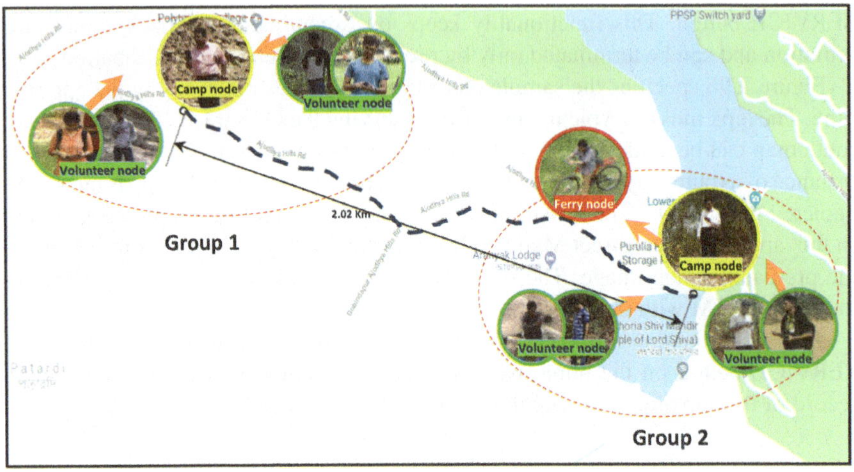

Fig. 2.10 Field trial conducted at Purulia, West Bengal, India

2.5.2 *Evaluating the Performance of Post-disaster Map Builder App*

The efficiency of the Post-disaster Map Builder app is estimated in this section depending on the measurements taken during the field trial carried out at Purulia district, West Bengal, India. Two parameters such as trace share ratio and average energy consumption are considered as the performance evaluation metrics. The representative graph of the trace share ratio over time is illustrated in Fig. 2.11a. It is noticeably evident from the graph that the trace share ratio increases gradually and reaches a steady-state after some time. It is also discovered from the outcomes that approximately 97.5% of GPS traces are conveyed to their destinations inside 2 h

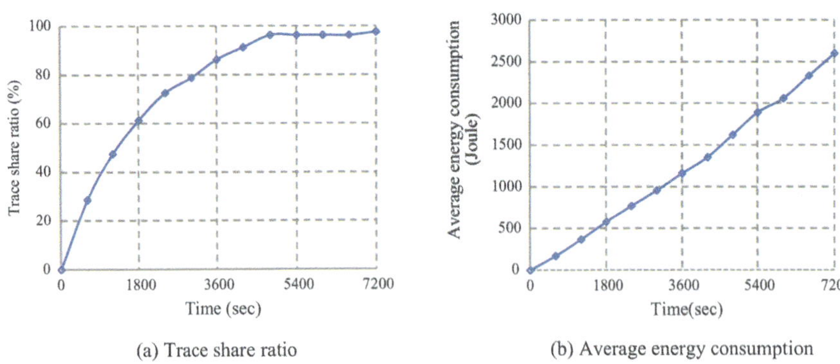

Fig. 2.11 Performance of Post-disaster Map Builder app

after the creation of such traces. Thus, the Post-disaster Map Builder app proves its competence in spreading GPS traces in the network.

The average energy requirement of every node for executing the app uninterruptedly is depicted in Fig. 2.11b. The figure illustrates that the energy requirement for executing the app for 2 h is roughly 2601 J, which is approximately 5.7% of the entire battery capacity. Hence, it can be predicted mathematically that all the nodes can run on battery for approximately 35 h assuming the device is only executing the Post-Disaster Map Builder app. So it can be forecasted that the nodes executing the Post-disaster Map Builder app can run on battery at least for an entire day after completely recharged, which is a crucial prerequisite for the post-disaster scenarios.

2.6 Conclusion

The proposed Post-Disaster Map Builder system comprises two units, namely *trace management* and *map inference*. In the *trace management* unit, the nodes gather trajectory traces with the help of the nodes' in-built GPS. Here, the nodes of a disaster-struck area exchange the gathered traces among themselves from time to time using the publish-subscribe messaging pattern. The *map inference* unit works in the mobile devices carried by the volunteers and makes route maps of the disaster-struck area using the traces gathered during the *trace management*. In the disaster-struck area, as the availability of stable power supply is ruled out and the handheld devices are limited in computing resources, the *map inference* unit is made lightweight, thereby implying energy saving. We measure the performance of our proposed system through simulation and testbed implementation. The results show that the devices are able to exchange traces 97% and above among themselves and it takes maximum 2 h to complete such exchange. It also shows that once the devices are re-energized they stay active for more than 24 h, which is good enough in a post-disaster environment. We also compare the simulation results with one competitor scheme [8], which shows that in spite of not using any high-end devices, our system's accuracy in generating route map is nearly at par with the competitor scheme.

The two other important post-disaster management services are situational awareness and resource management. Both of these services are invariably dependent on the existence of navigation support which is in place now. Thus, in the next chapter, two such services are elaborated.

References

1. Bhattacharjee, S., Roy, S., Das Bit, S.: Post-disaster map builder: Crowdsensed digital pedestrian map construction of the disaster affected areas through smartphone based DTN. Comput. Commun. **134**, 96–113 (2019). https://doi.org/10.1016/j.comcom.2018.11.010

2. Edelkamp, S., Schrödl, S.: Route Planning and Map Inference with Global Positioning Traces, pp. 128–151 (2003). https://doi.org/10.1007/3-540-36477-3_10
3. Schroedl, S., Wagstaff, K., Rogers, S., Langley, P., Wilson, C.: Mining GPS traces for map refinement. Data Min. Knowl. Discov. **9**, 59–87 (2004). https://doi.org/10.1023/B:DAMI.0000026904.74892.89
4. Cao, L., Krumm, J.: From GPS traces to a routable road map. In: Proceedings of the 17th ACM SIGSPATIAL International Conference on Advances in Geographic Information Systems—GIS '09, p. 3 (2009). https://doi.org/10.1145/1653771.1653776
5. Ahmed, M., Wenk, C.: Constructing Street Networks from GPS Trajectories (2012)
6. Wang, J., Rui, X., Song, X., Tan, X., Wang, C., Raghavan, V.: A novel approach for generating routable road maps from vehicle GPS traces. Int. J. Geogr. Inf. Sci. **29**, 69–91 (2015). https://doi.org/10.1080/13658816.2014.944527
7. Fathi, A., Krumm, J.: Detecting Road Intersections from GPS Traces (2010)
8. Karagiorgou, S., Pfoser, D.: On vehicle tracking data-based road network generation. In: Proceedings of the 20th International Conference on Advances in Geographic Information Systems—SIGSPATIAL '12, p. 89. ACM Press, New York, NY, USA (2012)
9. Liu, X., Biagioni, J., Eriksson, J., Wang, Y., Forman, G., Zhu, Y.: Mining large-scale, sparse GPS traces for map inference. In: Proceedings of the 18th ACM SIGKDD International Conference on Knowledge Discovery and Data Mining—KDD '12, p. 669. ACM Press, New York, NY, USA (2012)
10. Ercolano, J.M.: Pedestrian Disaster Preparedness and Emergency Management: White Paper for Executive Management. New York State Department of Transportation (2007)
11. Kasemsuppakorn, P., Karimi, H.A.: A pedestrian network construction algorithm based on multiple GPS traces. Transp. Res. Part C Emerg. Technol. **26**, 285–300 (2013). https://doi.org/10.1016/j.trc.2012.09.007
12. Blanke, U., Guldener, R., Feese, S., Troester, G.: Crowdsourced pedestrian map construction for short-term city-scale events. In: Proceedings of the First International Conference on IoT Urban Space (2014). https://doi.org/10.4108/icst.urb-iot.2014.257190
13. Qiu, J., Wang, R.: Inferring road maps from sparsely sampled GPS traces. J. Locat. Based Serv. **10**, 111–124 (2016). https://doi.org/10.1080/17489725.2016.1183053
14. Trono, E.M., Fujimoto, M., Suwa, H., Arakawa, Y., Yasumoto, K.: Generating pedestrian maps of disaster areas through ad-hoc deployment of computing resources across a DTN. Comput. Commun. **100**, 129–142 (2017). https://doi.org/10.1016/j.comcom.2016.12.003
15. Uddin, Y.S., Nicol, D.M., Abdelzaher, T.F., Kravets, R.H.: A post-disaster mobility model for delay tolerant networking. In: Proceedings of the Winter Simulation Conference, pp. 2785–2796 (2009). https://doi.org/10.1109/WSC.2009.5429249
16. Cao, Y., Sun, Z.: Routing in delay/disruption tolerant networks: a taxonomy, survey and challenges. IEEE Commun. Surv. Tutor. **15**, 654–677 (2013). https://doi.org/10.1109/SURV.2012.042512.00053
17. Zandbergen, P.A., Barbeau, S.J.: Positional accuracy of assisted GPS data from high-sensitivity GPS-enabled mobile phones. J. Navig. **64**, 381–399 (2011). https://doi.org/10.1017/S0373463311000051
18. Chazal, F., Chen, D., Guibas, L., Jiang, X., Sommer, C.: Data-driven trajectory smoothing. In: Proceedings of the 19th ACM SIGSPATIAL International Conference on Advances in Geographic Information Systems—GIS '11, p. 251 (2011). https://doi.org/10.1145/2093973.2094007
19. Wiedemann, C.: External evaluation of road networks. ISPRS Arch. **XXXIV**, 93–98 (2003)
20. Ben-Zur, L.: Trepn Power Profiler—Qualcomm Developer Network. https://developer.qualcomm.com/software/trepn-power-profiler

Chapter 3
Post-disaster Situational Awareness and Resource Management Services

The previous chapter provides the details of developing map-based navigation services in the affected areas over opportunistic networks. The solution to the two other important post-disaster services such as situational awareness and resource management are discussed in this chapter. As mentioned earlier both of these services are invariably dependent on the existence of navigation support. Precisely, without the map-based navigation services not only the volunteers are unable to reach the affected areas to gather situational information but also would fail to identify the exact routes toward resource inventories in such a scenario.

Situational awareness is a primary requirement for effective post-disaster response operations. In providing such service, the usage of opportunistic networks has two major limitations. One limitation is that dissemination of information over such networks is problematic due to the requirement of a reasonable quantity of technical knowledge from the users. The volunteers deployed by the response organizations might not possess such an amount of technical expertise. Moreover, due to security reasons the mobile handheld devices do not support data exchange without the consent of the user. However, the opportunistic networks-based application (app) development requires a data exchange between devices without user intervention. Such data exchange requires modifications of the pre-installed operating system (OS) of the device. The modification of the OS requires the root access of the device which increases the security vulnerability. Also, the apps developed for such modified OS are non-transferrable to other devices which hinders the circulation of the apps among the volunteers of various response agencies. Thus, in this chapter, the development of an Android app (Disaster Messenger) is presented, which offers post-disaster situational awareness service in absence of conventional network infrastructure without modifying the pre-installed OS of the device.

Apart from situational awareness, the other service is resource management which is also a vital part of disaster response operation. In this context, resources refer to materials like medicines, foodstuffs, clothes and other relief materials. In post-disaster scenarios, resources are managed in a decentralized manner. Camps set

S. Bhattacharjee et al., *Post-disaster Navigation and Allied Services over Opportunistic Networks*, Smart Innovation, Systems and Technologies 228,
https://doi.org/10.1007/978-981-16-1240-4_3

up by different disaster response agencies manage their resource inventories. Periodic dissemination of resource inventories generated at camps is vital to construct an integrated global inventory at each node which empowers effective search and retrieval of resource requests (queries). However, undermined network connectivity hinders the maintenance of such a uniform resource inventory and as a result, the effective search/retrieval of queries becomes difficult. In this chapter, a decentralized post-disaster resource management (DPDRM) scheme for offering such a service is described.

3.1 Literature Review

A literature review related to situational awareness and resource management services is provided below.

3.1.1 Situational Awareness

Situational awareness plays a vital role in effective post-disaster management. The primary objective of situational awareness is to gain accurate information rapidly such that the response efforts become productive. A substantial investigation has been performed regarding the accumulation of situational information in post-disaster scenarios. Such information gathering approaches can be largely divided into two classes such as information accumulation using traditional methods and social media-based information accumulation [1]. Some of the works from each category are reviewed below.

3.1.1.1 Traditional Information Accumulation

Traditional methods typically include phone calls, form-based information collection, direct observations, personal interviews, sensor data, and information accumulation through UAVs, etc. These methods have been commonly practiced by disaster response agencies to gain situational awareness.

In one such work [2], a wireless phone-based emergency response system (WIPER)is proposed. The system is developed using the notion of dynamic data-driven application systems (DDDAS) aiming to enhance situational awareness to the disaster response agencies in post-disaster scenarios. This system uses call data of cell phones from a geographic area. Cell phones are typically connected with single or multiple cell towers to make or receive calls. Thus, the positional data regarding all phones located in an area can be captured by tracking cell towers using triangulation methods. Analyzing the communication and mobility patterns of the population from disaster-affected areas, the disaster response agencies can assess the severity of the disaster and manage the response operation efficiently.

In another work [3], the authors describe the Human-Agent Collectives for Emergency Response (HAC-ER) toward improving situation awareness by accumulating information from various origins such as response teams, a fleet of UAVs and crowdsensing. In order to filter out inaccurate and unimportant data, the system incorporates a machine learning-based module Crowdscanner, which identifies the spots on the disaster-affected areas providing inaccurate information.

In another work [4] under this category, a WSN-based multiagent system for crisis management is proposed. The work focuses on tracing and data accumulation approaches to collect information about the situation and its evolvement.

In the other work [5] under this category, a UAV-cloud system is developed for recognizing the affected areas. A fleet of UAVs is used to accumulate visuals from the affected areas. In order to decrease the quantity of transmitted data, the system employs an image pre-processing mechanism on the UAVs based on the context information. The system incorporates several techniques such as video procurement, data scheduling, data offloading, data pre-processing and network analysis. Subsequently, the accumulated data is forwarded to the cloud servers for further processing.

All the works discussed above assume the presence of steady network connectivity which might be impractical in the disaster aftermath.

3.1.1.2 Social Media-Based Information Accumulation

Social media-based information accumulation includes mining post-disaster situational information from Twitter, Facebook, YouTube and other social media platforms [6]. Social media data allows the disaster response agencies to comprehend the severity of disaster and requirements of post-disaster scenarios and subsequently make decisions for effective post-disaster management.

In one such work [7], the existence of situational information is identified from tweets posted in the course of the Oklahoma Grassfires of April 2009 and the Red River Floods which took place during March and April 2009. Such information can help in gaining situational awareness in disaster aftermath. This work produces a framework of a software system that employs information extraction strategies to get situational information in emergency scenarios.

In another work [8], Twitter messages generated during natural disasters are analyzed. Authors present different relevant approaches such as burst detection, tweet filtering and classification, online clustering and geo-tagging for extracting situational information from tweets. The experimental results reveal that the extraction of suitable information from social media can help suitable establishments to make better decisions for disaster response by promoting their responsiveness to time-critical situations.

In the other work [9] under this category, a natural language processing (NLP)-based classification scheme is suggested which can identify tweets containing vital information from the tweets posted in the disaster aftermath. Trials over tweets associated with four different disaster events reveal that the proposed scheme recognizes situational awareness tweets with considerably higher precision.

Although social media helps disaster response agencies in gaining situational awareness, it also requires the existence of steady network connectivity in the affected areas which might not be realistic for all post-disaster scenarios. In addition to that, it is challenging to verify the authenticity of microblogs from social media. Hence, the possibility of spreading rumors in post-disaster scenarios through such platforms could not be ruled out.

3.1.2 Resource Management Services

One of the most prominent areas of research among the computer science community is content sharing. As this chapter presents a distributed content sharing notion concerning resource management, our literature review focuses on prevailing distributed content sharing schemes.

One modern scheme for searching content in opportunistic networks is suggested in [10]. The scheme is intended to limit the spreading of the queries. The authors propose four methods such as hop count, time to live, first identical response and global response count assessment in this scheme to control the unintentional dissemination of queries. It is evident from the experimental results that controlling the dissemination queries reduces network overhead significantly.

However, in another work under this category [11], the discover–predict–deliver (DPD) scheme is designed for opportunistic networks that predominantly emphasize content delivery. In this scheme, the locational information is used to restrict the unintentional duplication of queries in addition to the parameters used in [10]. This scheme also introduces two additional parameters such as split query lifespan limit and query distance limit to restrict the dissemination of queries. In order to forecast the future locations of the nodes, this scheme extracts the temporal and spatial regularity data from the node's movement characteristics. The performance of this scheme relies heavily on the precision of forecasting the mobility patterns of nodes. But, the scheme is not suitable for the conditions when the nodes demonstrate uneven mobility patterns and encounter outsiders.

In another work [12], a social-aware content and contact-based file search scheme (Cont2) is proposed in opportunistic networks. Unlike other schemes, both the interest and mobility characteristics of the nodes are considered in this scheme for content sharing. The network overhead is reduced in this scheme by employing a single copy routing protocol that exploits meetings and interests of the nodes to discover the required file. The scheme is normally intended for the distribution of bulky files through opportunistic networks.

In [13], an optimal scheme is presented for searching contents over opportunistic networks. In this work, the authors also study the effect of untrustworthy response routes on the resultant search performance. Moreover, the scheme identifies the key aspects which affect the optimum content search approach. However, this scheme presumes that both queries and responses follow the identical path.

As opposed to [11–13], in [14] another scheme is proposed which allows content sharing among the nodes in a situation where node mobility pattern remains unpredictable. This scheme requires steady network connectivity to obtain recommendations from social networks. The scheme considers unpredictable node movement and small stay duration of nodes in an area (e.g. visiting traveler locations, camping spots). It also supports content distribution depending on the interest of users and social networks-based recommendations. However, the scheme requires steady network connectivity to obtain recommendations from social networks.

The schemes discussed in [10–14] are intended for sharing high-volume data over the network. Although most of the existing schemes attempt to restrict the unintentional dissemination of the queries to decrease the network overhead, the schemes mentioned in [10–13] do not take into account the latency in obtaining a response upon sending a resource query. Further, the content sharing scheme in [11] achieves a fair amount of success when the mobility patterns of the nodes are regular. But, in the disaster aftermath, the nodes typically exhibit event-driven mobility patterns. Hence, symmetry in the mobility pattern could not be guaranteed. Another scheme discussed in [14] presumes the accessibility to steady network connectivity in post-disaster scenarios which is impractical.

The scheme described in [13] presumes both the queries and responses are forwarded through the identical routing path. However, due to frequent node mobility, the likelihood of the existence of a route from the query creation to response delivery is marginal. Moreover, the aforesaid schemes do not take into account the energy consumption of the nodes while evaluating performance. Therefore, the prevailing content sharing schemes remain unsuitable for resource management in the disaster aftermath where detection and delivery of the resource information are stringently time-bound.

3.2 Situational Awareness Service Through Disaster Messenger App

Numerous Android-based apps exist in the market for sharing files through Wi-Fi direct such as SHAREit [15] and BitTorrent Sync [16] etc. For instance, the SHAREit app facilitates the cross-platform (Android, Windows and iOS) transferring of files using Wi-Fi direct in the unicast mode of communication. There is another cross-platform app named BitTorrent Sync which supports the unicast mode of communication and transfers files using Wi-Fi direct. The primary limitation of BitTorrent Sync is that all the receiver nodes need to be added as a peer of the sender nodes which might be challenging in post-disaster scenarios.

The BitTorrent Sync typically employs peer-to-peer (P2P) protocols for transferring large files across devices. During file transfer, the app encrypts the files using

AES cipher and it does not keep any data on third-party servers. Thus, the files are safeguarded against identity breaches and malicious attacks. The aforesaid discussion reveals that the prevailing apps support data transfer in unicast mode only. Apps supporting multicasting and broadcasting are still absent in the market. However, the post-disaster situational awareness service requires the sharing of situational information in the form of text, image, video or recorded voice in absence of conventional network infrastructure in all three modes, such as unicast, multicast and broadcast. Hence, a comprehensive solution Disaster Messenger [17] in the form of an Android-based app is discussed in this section. The app allows the propagation of situational information by the volunteers through Wi-Fi direct in all three modes of communication. In order to facilitate the transmission and reception of data to/from numerous origins simultaneously, a new thread is built by the app for every transmission and reception.

3.2.1 *Design Modules*

The communication architecture of the Disaster Messenger app along with the functionalities of every module is discussed in detail in this section.

3.2.1.1 Communication Architecture of Disaster Messenger

The functionalities of the four modules of the app such as *authentication, user interaction, file manager* and *acknowledgment* are illustrated in Fig. 3.1.

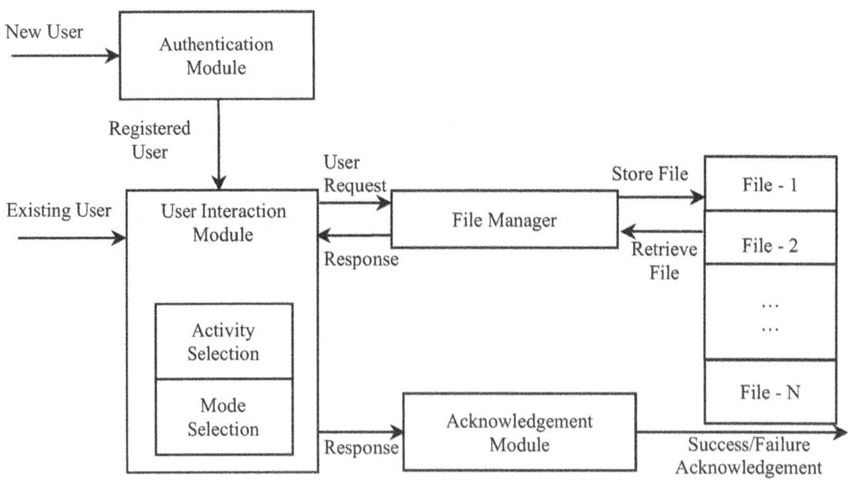

Fig. 3.1 Communication architecture of Disaster Messenger

The *authentication* module performs user validation for first-time users after the app is installed. Subsequently, the *user interaction* module takes care of user activities. This module contains a couple of sub-modules such as *activity selection* and *mode selection*. The *user interaction* module typically forwards the control to the *activity selection* and *mode selection* based on the preferences chosen by the users. The *activity selection* permits the users to select the send or receive activity. The *mode selection* permits to select the communication approaches such as unicast, multicast or broadcast, and in addition to that, the users can also pick the suitable recipient(s) as destination(s).

Subsequently, the transfer of control to the *file manager* is initiated which recovers or preserves files based on the user's request. Afterward, the control is reassigned to the *user interaction* module again. This module constructs a fresh thread for sending or receiving the request and handover the control to the *acknowledgment* module. This module primarily produces success or failure acknowledgment and forwards it to the *user interaction* module. Subsequently, the acknowledgment is visible to the users.

3.2.1.2 Flow Diagram

This section provides the flow diagram (Fig. 3.2) of data transfer activity. After the initiation of the Disaster Messenger app, the *authentication* module verifies the genuineness of first-time users and assign the control to the *user interaction* module. The selection of send or receive activities is performed in the *user interaction* module. Subsequently, the mode of communication is selected by the user. The control is then handed over to the *file manager*. This module recovers or preserves files from/to the auxiliary storage according to user request and forwards a reply to the *user interaction* module. After accepting the reply, the *user interaction* module constructs a fresh thread and validates the connection with the destination(s) or source as requested by the user. Finally, the file is transmitted or received. The *acknowledgment* module at that point acquires the control and forwards success or failure acknowledgment to the *user interaction* module.

3.2.1.3 Algorithm

The algorithm of Disaster Messenger is presented in this section. The algorithm presumes the existence of N devices moving into the network waiting to exchange situational information among themselves in one of the three modes, such as unicast, multicast or broadcast. When the sender(s)/receiver(s) arrives at the communication range, they construct a fresh set of threads for handling communication and hand over the control to the thread. The algorithm is executed at each device to exchange situational information.

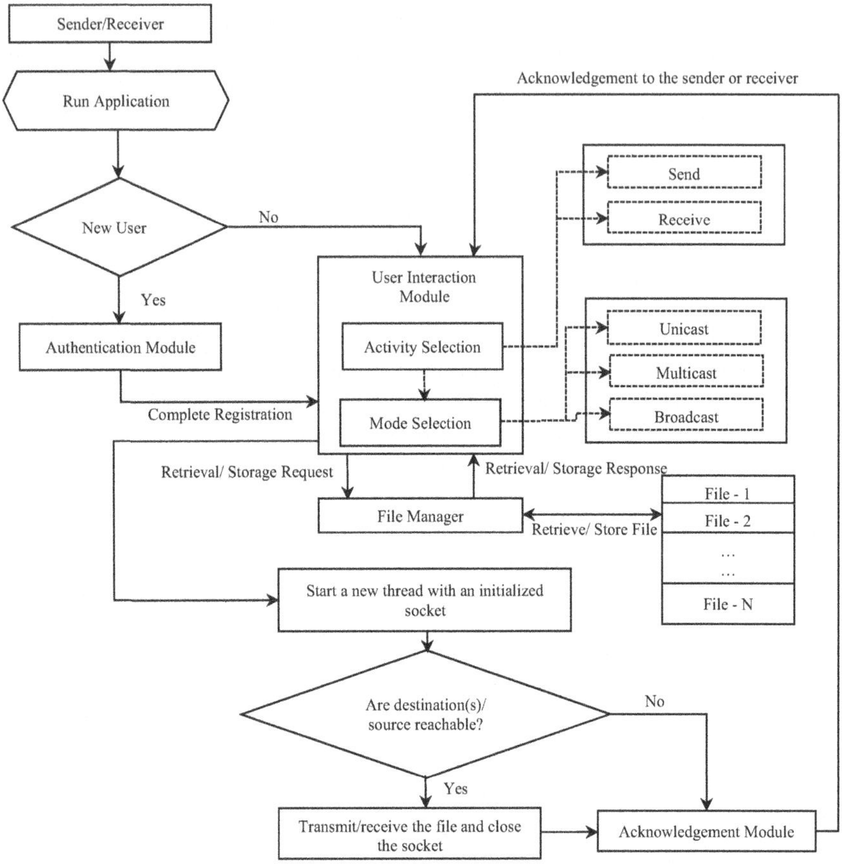

Fig. 3.2 Flow diagram of data transfer activity

The situational information to be disseminated amid numerous devices (nodes) is the inputs to the algorithm. The outputs of the algorithm are the messages to be sent and the received messages at the sender and receiver end, respectively.

Algorithm: Disaster Messenger
Input: Situational information to be shared among various devices
Output: Messages to be sent/received at the sender/receiver end
Algorithm Triggered: Upon request from the user

```
Begin
if (F_auth == False) then                        /* F_auth is device authentication flag*/
        set (D_id)                               /* D_id device id for the first-time user*/
end if
if (A_select == send) then                       /* A_select is the selected activity*/
  if (M_select == unicast) then                  /* M_select is selected mode*/
        D_id_dest ← store particular device id   /* D_id_dest stores the destination device id for unicast mode*/
elseif (M_select == multicast)
        D_id_dest ← select multiple device ids   /* list of destination device ids for multicast mode*/
else
        D_id_dest ← select all device ids        /* list of destination device ids for broadcast mode*/
end if
  send retrieve request to the file manager
choose file to be sent
  create (S_new)
  create (T_new)                                 /* S_new new socket*/
  start (T_new)                                  /* T_new new thread*/
  if (is_reachable (D_id_dest) == True) then
        while (F_ack == False) do                /* F_ack acknowledgement flag*/
              F_ack = check (Ack)
        end while
  end if
elseif (A_select == receive)
  send store request to the file manager
  create (S_new)
  create (T_new)
  start (T_new)
  while (F_ack== False) do
        F_ack = check (Ack)
  end while
end if
End
```

The running time complexity analysis of the algorithm indicates that the algorithm demonstrates the best-case complexity of O(1) and the worst-case complexity of O(n). The best case takes place when both the sender and the receiver(s) are inside the radio range of each other. In such a situation data transmission is instantaneous between sender and receiver(s). On the contrary, the worst-case takes place when the sender and receiver(s) are beyond the radio range and the sender repeatedly searches for the receiver(s).

3.2.2 *Performance Evaluation*

The performance of the Disaster Messenger app is analyzed in this section, through field trials under different scenarios.

3.2.2.1 Testbed Setup

We carried out a number of field experimentation at IIEST, Shibpur campus with the help of ten volunteers who carried smartphones/tablets installed with the Disaster Messenger app. We made three groups with three volunteers in each group and

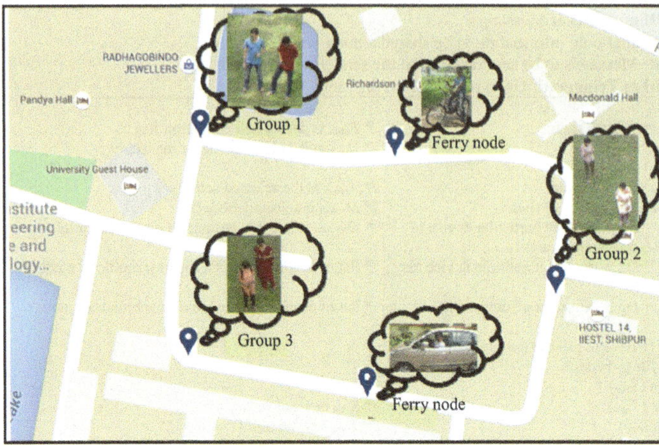

Fig. 3.3 Location of three volunteer groups at IIEST, Shibpur campus

deployed the groups in a manner that intergroup direct communication is not possible. In addition to this, one volunteer was made responsible to visit each of the groups at a time interval for collecting data from them. This roaming volunteer is typically known as *ferry nodes/data mules*. Whenever the *data mule* comes in contact with a group, it exchanges data with the group. A snapshot of the field trial shows (Fig. 3.3) the placement of three volunteer groups and one data mule.

To assess whether the proposed app works across various devices, during field trial smartphones and tablets produced by numerous vendors are used. The energy efficiency of our app (Disaster Messenger) is measured using an Android app PowerTutor [18]. Precisely, with the help of the Disaster Messenger app, one 83 MB video is transmitted/received from/to such devices 10 times and the energy consumption is recorded through the PowerTutor. Next, the arithmetic mean of all the recorded data including transmission and reception is calculated to get the average energy consumption. The detailed specifications of all the Android devices used in the experimentation are provided in Table 3.1. Figure 3.4 presents a few snapshots captured during field experimentation at IIEST, Shibpur.

3.2.2.2 Results and Discussion

Primarily we have conducted an experiment to assess the variation of average data transfer rate in all three modes, such as unicast, multicast and broadcast by varying the distance between the sender and the receiver. Figure 3.5 plots the change of data transfer rate. The results confirm that the data transfer rate is significantly reduced while the distance between the sender and the receiver increases in all three modes.

Subsequently, we have conducted three sets of experiments that compare the results with two competitor apps such as SHAREit and BitTorrent Sync. All these

Table 3.1 Device specifications for testbed implementation

Device name	Device category	Memory capacity (MB)	Storage capacity (GB)	Battery capacity (mAh)	Wi-Fi direct range outdoor (m)
Samsung Galaxy S Duos 2	Phone	768	4	1500	80.5
Samsung Galaxy J5	Phone	2048	16	3300	106
Moto E (1st Gen)	Phone	1024	4	1980	47.5
Samsung Galaxy TabS-1	Tablet PC	3072	32	5870	57
Samsung Galaxy TabS-2	Tablet PC	3072	32	7900	106
Moto G2	Phone	1024	8	2070	47.5

Fig. 3.4 Snapshots of field trials conducted at IIEST, Shibpur

experiments are conducted for the unicast mode of communication. The SHAREit and BitTorrent Sync are two widely accepted Android apps that offer data transferring in absence of conventional network infrastructure, which is a primary requirement for communication in the disaster aftermath. Due to the absence of suitable rival apps that offers multicast and broadcast modes in absence of conventional network infrastructure, the aforesaid apps are chosen for evaluating the performance of Disaster Messenger.

Fig. 3.5 Average data transfer rates with distance in three different modes

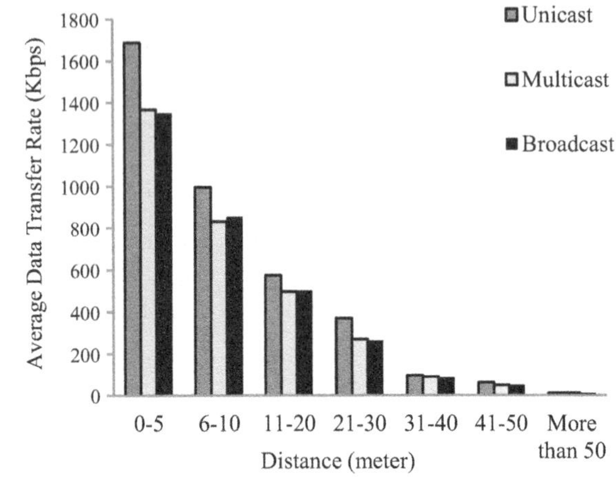

In the first set of experiments, three competitor-apps Disaster Messenger, SHAREit and BitTorrent Sync are evaluated in terms of average data transfer rate over the sender-to-receiver distance in unicast mode. The results are plotted in Fig. 3.6a. Particularly, we notice that the data transfer rate of Disaster Messenger is 8.75%

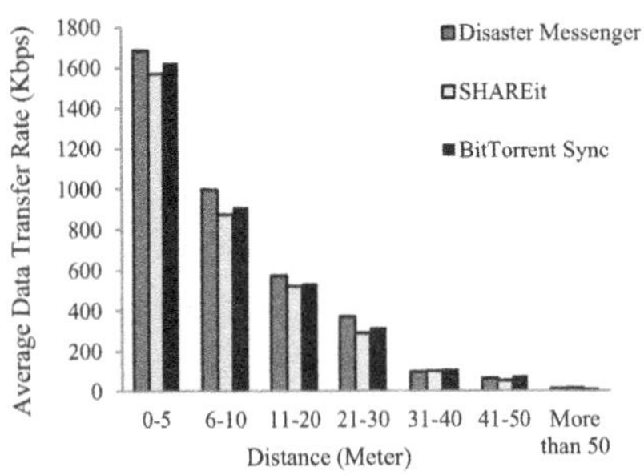

(a) Average data transfer rates with distance in unicast mode

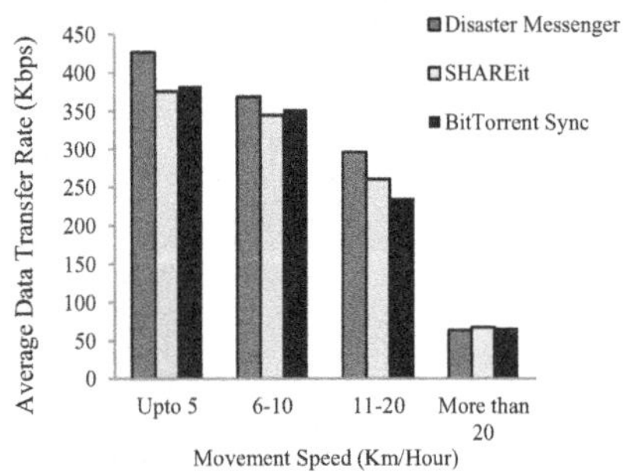

(b) Average data transfer rates with movement speed in unicast mode

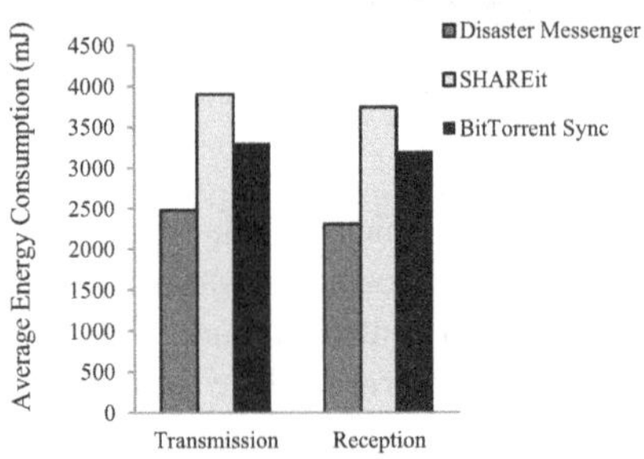

(c) Average energy consumption in unicast mode

Fig. 3.6 Comparative performance of Disaster Messenger

and 2.81% higher than its two such competitors SHAREit and BitTorrent Sync, respectively.

Similarly, in the second set of experiments, the performance of the three competitor apps, including ours, is compared in terms of the movement speed of the nodes considering the unicast mode of data transfer. The experimental outcomes results are depicted in Fig. 3.6b. Here also, we observe that the data transfer rate of Disaster Messenger is approximately 7.1% and 9.2% higher than SHAREit and BitTorrent Sync, respectively.

Thus, the above two sets of results establish the dominance of Disaster Messenger with respect to the average data transfer rate in unicast mode over its competitors.

On the other hand, in the third set of experiments, the comparison is made in terms of average energy consumption during data communication in unicast mode. The results are plotted in Fig. 3.6c. It is observed from the results that the Disaster Messenger requires approximately 37% and 33% lesser transmission energy than SHAREit and BitTorrent Sync, respectively. In the case of reception energy, these values are 38.4 and 36.4% less.

Therefore, regarding both the parameters energy consumption and data transfer rate, Disaster Messenger performs better than SHAREit and BitTorrent Sync.

Disaster Messenger typically offers a user-friendly interface without demanding any kind of technical proficiency from end-users. The aforesaid benefits make it distinctive and appropriate for use in post-disaster scenarios. Thus, Disaster Messenger offers situational awareness service by exchanging various types of situational data including multimedia in unicast communication. The other important service in disaster aftermath is resource management which is elaborated in the following section.

3.3 Resource Management Service Through DPDRM Scheme

Another integral part of disaster response is resource management. Our interactions with various disaster response NGOs in India such as *Doctors For You*[1] (DFY) and *Society for Promotion of Appropriate Development Efforts*[2] (SPADE) disclose that in developing countries like India post-disaster resource management is manual and paper-based procedure at camps. This manual resource management process reduces volunteer awareness regarding resource inventories and accordingly hinders the effective search and retrieval of resources. Moreover, a considerable amount of time and toil needs to be devoted to sustain a near-accurate, global inventory of

---

[1]Doctors For You (DFY) is a pan India humanitarian organization with international presence and is working in various disaster hit zones since 2007. Website: http://doctorsforyou.org/

[2]Society for Promotion of Appropriate Development Efforts (SPADE) is a voluntary non-governmental organization in eastern India formed in 1994. Website: http://www.spadeindia.org/

resources manually. Apart from that, manual resource management typically necessitates direct human intervention which makes it error-prone. These weaknesses of manual resource management result in the inconsistent circulation of life-saving resources causing unsolicited human casualties. The above-mentioned weaknesses of manual resource management are the primary inspiration toward the development of an automated and low-latency scheme, decentralized post-disaster resource management (DPDRM) [19], over opportunistic networks which can cater to the necessities of post-disaster scenarios. The DPDRM scheme attains its goals by spreading resource inventory periodically amid the *volunteer nodes* connected with analogous groups and by handling queries regarding resources produced by *volunteer nodes*.

3.3.1 System Architecture

The detailed system architecture of DPDRM accompanied by assumptions is discussed in this section. It has been observed that in disaster aftermath diverse response agencies such as Medical Support Group, Relief Group, Water and Sanitation Group, etc. set up camps nearby the disaster-struck regions. The camps are primarily responsible for storing, distributing and managing resources (e.g. medicines, foodstuff, etc.) supplied to them in a decentralized manner. Each camp typically maintains a resource inventory that changes over time. Hence, the information available in the resource inventories has a certain lifespan and becomes obsolete afterward. Thus, decentralized post-disaster resource management operates under strict time constraints. A command and control center is established at a faraway position from the affected area, for allocating and transporting resources to various camps using supply vehicles as per resource requests from such camps. The command and control center does not supervise the local resource circulation procedure carried out by the camps. The system architecture of the DPDRM scheme is demonstrated in Fig. 3.7. It uses Ge-PRoPHET, a PRoPHET-like routing protocol to be discussed in Chap. 5 for data dissemination. The scheme employs the publish-subscribe mechanism [20].

The resource inventories are typically produced by the camps. The *camp nodes* broadcast (publish) those inventories periodically amid the *volunteer nodes* connected with the designated response groups. In post-disaster scenarios, the term designated response groups indicate those groups whose functional area is similar to the inventory publisher. The *volunteer nodes* of the respective response groups allow (subscribe) to keep the broadcast inventories inside their handheld devices (smartphone). A real example may be the volunteers of the medical-support group act as potential subscribers of the inventories such as medicine which is published by the camps dealing with medical items. All the nodes (e.g. camp, volunteer) maintain a catalog of records (*inventory index*) that are kept at the nodes.

The queries (resource requests) are typically generated by the *volunteer nodes*. Upon generating queries, the *volunteer nodes*, primarily search their local storage for the required information. In case of unavailability of the required information,

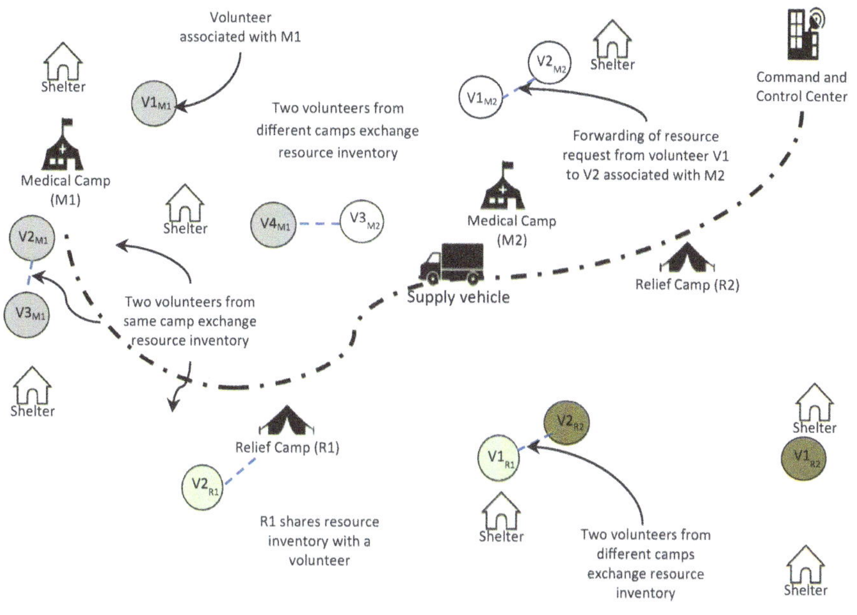

Fig. 3.7 System architecture of decentralized post-disaster resource management (DPDRM) scheme

the queries are dispatched to the member nodes of the designated response groups. A catalog of the queries (*query index*) generated/received is retained at each node. Once two nodes come inside the radio range of each other they interchange their *inventory indexes* and *query indexes* with other nodes. This interchange allows nodes to determine which resource inventories and queries are available with the encountered node(s). Accordingly, the transfer of resource inventories and queries takes place.

The opportunistic networks considered here consist of two types of nodes: static and mobile. Here, the *camps* and *shelters* belong to the category of static nodes, whereas *supply vehicles* and *volunteers* are mobile nodes. The *volunteers* demonstrate the post-disaster movement (PDM) model [21]. The nodes are homogeneous in terms of initial energy and energy consumption for transmission/reception. The *volunteer nodes* generate queries at random at 1–2 h gap. The *camp nodes* update inventories once a day. The network architecture is considered as a *three-tier* architecture which will be elaborated in Chap. 4.

3.3.2 *Proposed DPDRM Communication Architecture*

The primary activities of various modules of the proposed DPDRM scheme as well as the interconnections among them are discussed in this section. The communication architecture of DPDRM is depicted in Fig. 3.8. The two modules, *inventory sharing* and *resource discovery* entirely manage the whole scheme.

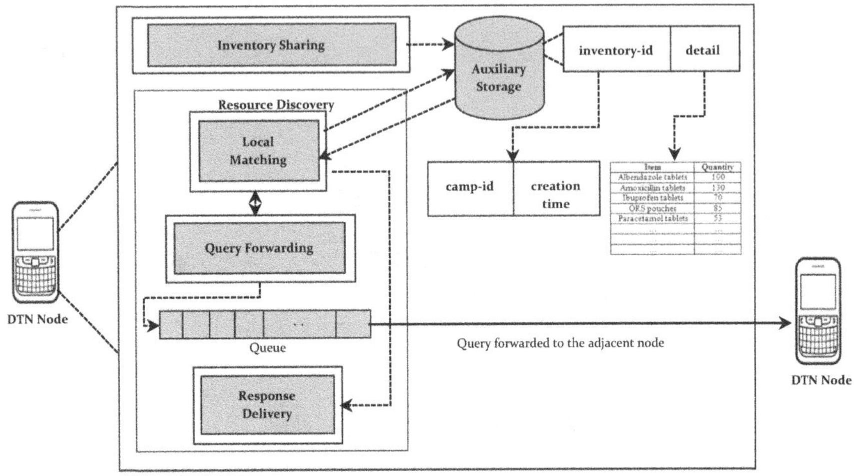

Fig. 3.8 Communication architecture of the proposed DPDRM scheme

The *inventory sharing* module facilitates the dissemination of resource inventories among the *volunteer nodes* linked with a specific group. This group-based dissemination (*groupcasting*) of resource inventories enables proficient detection of resources by minimizing resource discovery time.

The *resource discovery* module facilitates the detection of required resource information and distribution of the same to the initiator node over opportunistic networks. This module is divided into two, such as *query forwarding* and *response delivery* submodules. In the disaster aftermath, the resource inventories are typically disseminated amid a set of *volunteer nodes*. Hence, group-based multicasting (*groupcasting*) of resource inventories is a key requirement in such scenarios. Moreover, it is almost impossible to recognize a node beforehand which can respond to a query. Therefore, *groupcasting* remains indispensable for query dissemination. Existing opportunistic network routing protocols do not inherently support *groupcasting*. Therefore, we have enhanced the PRoPHET protocol to support *groupcasting*. The detailed activities of each module are discussed in the following sections.

3.3.2.1 Inventory Sharing

The *groupcasting* of resource inventories is accomplished by the *inventory sharing* module with the help of group encounter-based PRoPHET (Ge-PRoPHET) routing protocol. The Ge-PRoPHET protocol will be discussed in detail in Chap. 5. It has been observed that the quantity of *camp nodes* is rarer in comparison with the *volunteer nodes* in the affected areas. Therefore, searching for resource inventories from those inadequate numbers of *camp nodes* over opportunistic networks becomes time-taking. In addition to that, in post-disaster situations, stringent time restrictions are imposed on resource detection. Hence, without relying only on *camp nodes* as the sole providers of resource inventories in the network, we attempt to raise the quantity of resource inventory provider nodes. Our aim of increasing the quantity of resource inventory provider nodes is fulfilled by the *inventory sharing* module through the dissemination of inventories amid the *volunteer nodes* of designated response groups. Thus, the resource discovery time is also considerably reduced.

We introduce a utility function in this module termed inventory forwarding predictability (P_{f_inv}), which can be estimated from group delivery predictability (GDP). The GDPs of different nodes in the network are stored in a table (GDP table) in each node. The GDP table keeps track of the delivery predictability of nodes with all other groups deployed in the affected area. A detailed discussion on GDP can be found in Chap. 5. The primary focus of P_{f_inv} is to identify an appropriate forwarder node for a resource inventory.

In addition to that, the *inventory sharing* module also reduces the possibility of disseminating outdated resource inventories. It offers a better chance of forwarding to the resource inventories which has a higher likelihood of becoming outdated. This objective is fulfilled by the combination of time-to-live predictability (TTLP) of that resource inventory and forwarding predictability. The estimation of the TTLP of resource inventories is elaborated as follows.

The time-to-live (TTL) of a resource inventory (ttl_{inv}) specifies the residual lifespan of that inventory before becoming outdated. A pre-defined TTL (ttl_{inv_def}) of 12 h and a creation timestamp ($t_{inv\_generate}$) is allocated to each resource inventories when they are created. Therefore, both the ttl_{inv} and ttl_{inv_def} retains identical values at the time of inventory creation. The ttl_{inv} is progressively reduced as time passes. At some particular time ($t_{current}$), ttl_{inv} is calculated as follows:

$$ttl_{inv} = (ttl_{inv\_def} - (t_{current} - t_{inv\_gennerate})) \tag{3.1}$$

TTLP of a resource inventory ($P_{ttl\_inv} \in [0, 1]$) signifies the likelihood of the existence of the resource inventory in near future and is estimated as follows:

$$P_{ttl\_inv} = \frac{ttl_{inv}}{ttl_{inv\_def}} \tag{3.2}$$

A higher P_{ttl_inv} value signifies a better chance that a resource inventory will remain updated (survive) in near future. In this module a resource inventory acquires the depending upon the utility function forwarding predictability ($P_{f\_inv} \in [0, 1]$) which is calculated as follows:

$$P_{f\_inv} = P_{GDP\_contacted} \times (1 - P_{ttl\_inv}) \tag{3.3}$$

In Eq. 3.3, the GDP of an encountered node is symbolized by $P_{GDP\_contacted}$. The $P_{GDP\_contacted}$ is determined using the well-known formula of the theory of probability, $P(X \cup Y) = P(X)P(Y)$ where X and Y are independent events. Likewise, in our application domain, P_{f_inv} relies on two probabilities, viz., $P_{GDP\_contacted}$ and P_{ttl_inv}, both of which are independent. Thus, the product operation is considered during the computation of P_{f_inv}. The P_{f_inv} offers a better opportunity for transmission to resource inventories with lower TTLP and vice versa.

Let us consider that INV_1 and INV_2 are two inventories of two camps maintained by a *volunteer node* V_A. The corresponding P_{ttl_inv} values of INV_1 and INV_2 are 0.7 and 0.4 destined for G_2 and G_3 groups, respectively. Now, let us assume that the V_A comes in contact with V_B, another *volunteer* whose GDP is higher than the V_A. If the V_B's $P_{GDP\_contacted}$ are 0.9 and 0.6 for the G_2 and G_3, respectively, as per Eq. 3.3, the P_{f_inv} are 0.27 and 0.36 corresponding to INV_1 and INV_2. Hence, it is evident that the chance of forwarding is higher for INV_2 which has a lower TTLP. The *inventory sharing* algorithm is as follows:

Algorithm: Inventory sharing
Input: Resource inventories (*inv[ ]*) to be shared
Output: Identifying the resource inventories to be forwarded to an encountered node
Algorithm Triggered: Upon contact with other nodes

```
Begin
for each inv_id in inv[i]
    fetch P_GDP_contacted                 /* by searching the GDP table of the encountered node*/
    fetch P_GDP_source                    /* by searching the GDP table of source node */
if (P_GDP_contacted > P_GDP_source) then  /* encountered node has higher GDP */
        compute P_f_inv                   /* using Equation (3.3) */
inv_table ← (inv_id, P_f_inv)             /* insert 2-tuple in the inv_table , sorted by P_f_inv */
i ++                                      /* incrementing loop counter */
end if
end for
while (inv_table is not empty) do
    forward inventory to encountered node
end while
End
```

3.3.2.2 Resource Discovery

The primary purpose of the *resource discovery* module is to spread queries and delivering the related responses to the creator node. This module is present in all the nodes including camp and volunteer nodes and is triggered instantly upon creation/arrival of a fresh query at the nodes. In case the required resource information is unavailable/outdated in the resource inventories (determined by ttl_{inv}) retained by the creator node, the spreading of the query to the encountered nodes is initiated by this module for information retrieval. In addition to that, this module stops the query spreading and delivers the accumulated information regarding the query to the *response delivery* sub-module in case the information concerning a query is present in the resource inventories ($ttl_{inv} > 0$) retained by a node. The *response delivery* sub-module is also accountable for delivering the response to the query creator node. The address of the creator node is obtained from the query. The resource inventories to be disseminated among different nodes are the inputs to the resource discovery algorithm. The resource inventories to be forwarded to an encountered node are the outputs from the same algorithm.

Query forwarding

The *query forwarding* sub-module is invoked upon the creation or arrival of a query at a node. The primary purpose of this sub-module is to regulate the spreading and duplication of the queries. Primarily, this sub-module tries to procure the necessary information from the auxiliary storage (*local matching*). The *local matching* permits each node to explore resource inventories stored in their auxiliary storage for required resource information. In case the necessary data is obtained by *local matching*, it is sent to the *response delivery* sub-module. Otherwise, the query is disseminated to the encountered nodes. The forwarding nodes are either associated with the identical response group or with dissimilar response groups with higher encounter possibilities. In order to determine the likelihood of fetching the required information regarding a resource in the auxiliary storage of a node, the response predictability ($P_{rsp} \in [0, 1]$) is introduced and is defined as follows:

$$P_{rsp} = \begin{cases} 1 \text{ if } ttl_{inv} > 0 \text{ and resource is available in the resource inventory} \\ 0 \text{ otherwise} \end{cases} \quad (3.4)$$

It can be observed from Eq. 3.4 that the P_{rsp} is set to 0 in case of outdated ($ttl_{inv} = 0$) resource inventory or unavailability of requested resource information.

Identifying a node that will respond to a query is almost impossible. Hence, spreading queries using unicast communication does not produce the anticipated result as the actual destination of the queries could not be determined in advance. Moreover, spreading queries using multicast communication considerably decreases query replication causing a substantial reduction in network overhead. So, the queries need to be disseminated using multicast communication. Thus, the *query forwarding* sub-module utilizes the Ge-PRoPHET protocol for query spreading. The Ge-PRoPHET protocol is an enhanced version of the well-known PRoPHET routing protocol which facilitates the *groupcasting* in opportunistic networks.

The TTL of a query (ttl_{qry}) signifies the residual lifespan of that query. Each query is allocated a default TTL (ttl_{qry_def}) of 12 h and a creation timestamp ($t_{qrv\_generate}$) upon generation. The ttl_{qry} is reduced gradually over time. At some particular instant ($t_{current}$) it is calculated as follows:

$$ttl_{qry} = \begin{cases} 0 & \text{if } P_{rsp} = 1 \\ \left(ttl_{qry\_def} - \left(t_{current} - t_{qry\_gennerate}\right)\right) & \text{otherwise} \end{cases} \quad (3.5)$$

TTLP of a query ($P_{ttl\_qry} \in [0, 1]$) implies the likelihood of survival of the query in near future and is estimated similarly like Eq. 3.2.

$$P_{ttl\_qry} = \frac{ttl_{qry}}{ttl_{qry\_def}} \quad (3.6)$$

In our application domain, query forwarding predictability (P_{f_query}) relies on $P_{GDP\_contacted}$, P_{rsp} and P_{ttl_qry}. Therefore, applying the laws of probability P_{f_query} is estimated as follows:

$$P_{f\_query} = \begin{cases} 0 & \text{if } P_{ttl\_qry} = 0 \\ P_{GDP\_contacted} \times \left(1 - P_{ttl\_qry}\right) \times \left(1 - P_{rsp}\right) & \text{otherwise} \end{cases} \quad (3.7)$$

It can be observed from the Eq. (3.7) that P_{f_query} provides the opportunity of transmission to the queries having a reasonably lower residual lifespan (ttl_{qry}). The P_{rsp} and P_{ttl_qry} manage the unwanted replication and termination of queries, respectively. For instance, if required resource information is fetched from the auxiliary storage of a node by *local matching*, P_{rsp} is set to 1 causing P_{f_query} is set to 0. This indicates the termination of the query. Likewise, in the case of expiry of the residual lifespan of a query ($P_{ttl\_qry} = 0$) the query needs to be terminated. Such control over replication, as well as dissemination of queries, significantly curtails the energy consumption due to query transmission.

Response delivery

The *response delivery* sub-module is accountable for delivering the responses to the query creator node. It typically accepts the information from the *local matching* sub-module. Identical to the *query forwarding* this sub-module also uses the Ge-PRoPHET protocol for transmitting responses. The delivery of the response can be managed by merging the TTLP of the response ($P_{ttl\_rsp} \in [0, 1]$) along with $P_{GDP\_contacted}$. The TTL of a response (ttl_{rsp}) signifies the residual lifespan of that response. In case it does not arrive at the creator node within 12 h of query creation, then it is anticipated that the requested resource is unavailable at the *camp nodes* of the affected area. Thus, the ttl_{rsp} relies on the ttl_{qry}. At a certain time ($t_{current}$), ttl_{rsp} is computed as follows:

$$ttl_{rsp} = ttl_{qry\_def} - (t_{current} - t_{qry\_gennerate}) \quad (3.8)$$

The TTLP of any response (P_{ttl_rsp}) implies the likelihood of survival of that response in near future and is estimated using the equation as follows:

$$P_{ttl\_rsp} = \frac{ttl_{rsp}}{ttl_{qry\_def}} \quad (3.9)$$

In this sub-module, response forwarding predictability (P_{f_rsp}) is calculated using the equation as follows:

$$P_{f\_rsp} = P_{GDP\_contacted} \times (1 - P_{ttl\_rsp}) \quad (3.10)$$

Equation 3.10 indicates P_{f_rsp} is designed in such a way that the responses with lower ttl_{rsp} receive higher forwarding opportunities. The entire algorithm of *resource discovery* is described below where the queries to be shared among different nodes are the inputs to the algorithm. The outputs of the algorithm identify suitable encountered nodes as potential forwarders for the responses of such queries. The *resource discovery* algorithm is given below:

```
Algorithm: Resource discovery
Input: Queries (qry[ ]) to be shared
Output: Responses to be forwarded to an encountered node
Algorithm Triggered: Upon generation/reception of a query
Begin
for each qry_id in qry[j]
  compute P_rsp                              /* compute response predictability using Equation 3.4 */
  if (P_rsp == 1) then                       /* requested information is available at the source node*/
rsp ← fetched resource using local matching
    call query_forwarding(qry_id, P_rsp)     /* terminating query using Equation 3.7*/
    call response_delivery(qry_id, rsp)      /* triggered upon contact with encountered node */
  else
    if (P_GDP_contacted > P_GDP_source) then
      call query_forwarding(qry_id, P_rsp)   /* forward query to the encountered node */
end if
end if
j++                                          /* incrementing loop counter */
end for
/*Function*/
response_delivery(qry_id, rsp)
compute P_GDP_contacted                      /* GDP of the encountered and source node with destination
compute P_GDP_source                         group of the response */
if (P_GDP_contacted > P_GDP_source) then     /* encountered node has higher GDP */
  compute Pf_rsp                             /* using Equation 3.10 */
  if (Pf_rsp > 0)
rsp_table ← (Pf_rsp, rsp)                    /*store Pf_rsp and rsp in rsp_table in sorted order */
end if
end if
while (rsp_table is not empty) do
  forward response to the encountered node
end while

/*Function */
query_forwarding (qry_id, P_rsp)
tmp_table ← (qry_id, P_rsp)                  /* store 2-tuple in tmp_table */
compute Pf_query                             /* using Equation 3.7 */
if (Pf_query > 0) then
qry_table ← (qry_id, Pf_query)               /* store Pf_query and qry_id in qry_table sorted by Pf_query */
end if
while (qry_table is not empty) do
  forward query to the encountered node
end while
End
```

3.3.3 Overhead Analysis

In this section, the DPDRM scheme is evaluated through theoretical analysis based on communication, computation and storage overheads. During overhead estimation, energy consumption is used to assess communication and computation overheads. The storage demand of each node for running both the algorithms is used to assess storage overhead. Although in reality the nodes are originated from different

Table 3.2 Energy consumption for relevant algorithmic construct (ARM Cortex A7 1200 MHz)

Algorithmic constructs	Energy consumption/instruction (pJ)
LOAD	141
COMPARE	50
STORE	161.6
ADD	57
MUL	45
MOV	43
SUB	57
DIV	46

manufacturers and possess diverse hardware configurations, for the sake of accessibility of data, ARM Cortex A7 microprocessor specification is considered for evaluation of overheads. The required energy consumptions [22–24] to realize pertinent algorithmic constructs are provided in Table 3.2. The overhead analysis is also done for one existing competitor scheme DPD (discover–predict–deliver) [11] for comparison.

3.3.3.1 Communication Overhead

This overhead is typically caused by the transmission and reception of resource inventories, queries and responses. The communication energy is calculated by taking the product of the power consumption for transmission or reception along with the duration of transmission/reception. We assume the total quantity of nodes in a network is n at an instant of time. We further assume that, out of total n nodes, the quantity of camp nodes is c. Hence, the quantity of *volunteer nodes* is (n-c). We utilize the Samsung Galaxy Grand 2 device specification for overhead analysis.

Transmission: As per the device description, power consumption during data transmission at the uppermost speed (750 KBps) is 1526 mW. So energy requirement for sending single data bit is: 0.25 μJ.

Energy consumption of **DPDRM** for transmission is due to the following two modules:

(a) *Inventory sharing:* It is assumed that the highest capacity of each resource inventory is 16 KB. Thus, the module needs at least 16 KB data for sending a single inventory, $8n$-bit data is required for the address of the encountered nodes and $24n$-bit data is required for transmitting the GDP table. The GDP table contains a couple of entities such as the address of the node and GDP. The address field requires $8n$-bit data for sending node addresses and the GDP field requires $16n$-bit data for GDP. Thus, the total storage requirement for the GDP table is $24n$-bit. Hence, the overall energy requirement for inventory sending is: $0.25(32n + 131072c)$ μJ.

(b) *Resource discovery:* This module necessitates at most 1 KB data for sending each query, 146 B data for each response, $8n$-bit for data for node address and $24n$-bit data for GDP table. Hence, the overall energy requirement for sending both query and response is:$0.25(32n + 9360)$ µJ.

Further, the transmission energy consumption of **DPD** is derived as follows:

The scheme also necessitates at most 1 KB data for sending a single query, 146 KB data for a single response (assuming the usual space requirement for each content is 500 KB), $16n$-bit data for storing GPS trace (8-bit each for latitude and longitude) and $8n$-bit data for addresses of the encountered nodes. Hence, the overall energy requirement for sending both query and response is: $0.25(24n + 9360)$ µJ.

Reception: It has been observed that the power consumed for the reception of data at a speed of 750 KBps is 1479 mW. Thus, the energy requirement for the reception of a single data bit is: 0.24 µJ.

Energy consumption of **DPDRM** for the reception is also due to the following two modules:

(a) *Inventory sharing*: During the reception, this module necessitates an identical amount of data like a transmission for resource inventory and GDP table. Hence, the overall energy requirement for the reception of resource inventory is: $0.24(24n + 131072c)$ µJ.

(b) *Resource discovery*: In the case of reception, this module necessitates an identical amount of data like a transmission for each query, response and GDP table. Hence, the overall energy requirement for receiving both query and response is:$0.24(24n + 9360)$ µJ.

Also, the energy consumption of **DPD** for the reception is derived as follows.

During the reception, the DPD scheme also needs an identical amount of data like a transmission for each query, response and GPS trace of the nodes. Hence, the overall energy requirement for the reception of both query and response is: $0.24(16n + 9360)$ µJ.

3.3.3.2 Computation Overhead

In order to calculate computation overhead, each high-level instruction is converted to the corresponding low-level instructions. Let us consider the total quantity of nodes is n, the total quantity of resource inventories is k and the total quantity of queries generated by all nodes is p. We assume each query will get a response. Hence the total number of responses is also k.

DPDRM: Computation overhead is calculated in terms of energy. It is due to the following two modules:

(a) *Inventory sharing*: The total quantity of low-level instructions to be executed by this module is: $(32nk + 18)$.

Energy required $= 3699.2nk + 2340.4$ pJ.

(b) *Resource discovery*: The total quantity of low-level instructions to be executed by this module is:$(179np + 104)$.

Energy required $= (17124.4np + 4736.8)$.

DPD scheme: Computation overhead in terms of energy is derived as follows:

The overall quantity of instructions to be executed is: $(57n^2 + 78n + 186)$.

Energy required $= (1379.2n^2 + 2172.6n + 3175)$.

3.3.3.3 Storage Overhead

This overhead theoretically estimates the maximum secondary storage space requirements for executing a scheme at each node.

DPDRM: In the DPDRM scheme, it is assumed that the n nodes retain k resource inventories. Each inventory requires a storage space of at most 16 KB. So, the overall storage necessity for k resource inventories in each node is $(16 \times 1024 \times 8 \times k) = 131072k$ bits. Moreover, it is also assumed that p queries are created from each of the n nodes where each query requires at most 1 KB storage space. Thus, the overall storage necessity for p queries in each node is $(1 \times 1024 \times 8 \times p \times n) = 8192pn$ bits. Hence, the total storage requirement for each node in the DPDRM scheme is $(8192 \times p \times n + 131072 \times k)$ bits.

DPD: Like the DPDRM scheme, the overall storage necessity in this scheme for storing k resource inventories in each node is $(16 \times 1024 \times 8 \times k)$ bits. In addition to that, in DPD each node requires an additional 16-bit storage space each for storing GPS traces of other nodes. Hence, the storage need for GPS traces is $(16 \times n)$ bits. Also, the storage necessity for p queries in each node is $(1 \times 1024 \times 8 \times p \times n)$ bits. Therefore, the overall storage space requirement for every *camp node* in the DPD scheme is: $(8192 \times p \times n + 16 \times n + 131072 \times k)$ bits.

Table 3.3 presents comparative communication, computation and storage overhead for both the contending schemes. The communication and computation overhead is measured in terms of energy.

3.3.3.4 Illustrative Example

Let us assume a post-disaster scenario having 130 nodes (7 *camp nodes* and 123 *volunteer nodes*). It is further assumed that each *volunteer node* creates three queries in each 12 h interval, and each *camp node* produces fresh resource inventories twice a day. The overall transmission and reception after 12 h are 300 (180 for inventory sharing and 120 for resource discovery for DPDRM) each for both the schemes.

Table 3.3 Comparative energy and storage overhead of DPDRM scheme

Scheme		Communication energy (μJ)		Computation energy (pJ)	Storage overhead (bits)
		Transmission energy	Reception energy		
DPDRM	Inventory sharing	$0.25(32n + 131072c)$	$0.24(24n + 131072c)$	$3699.2nk + 2340.4$	$8192pn + 131072k$
	Resource discovery	$0.25(32n + 9360)$	$0.24(24n + 9360)$	$17124.4np + 4736.8$	
DPD		$0.25(24n + 9360)$	$0.24(16n + 9360)$	$1379.2n^2 + 2172.6n + 3175$	$8192pn + 16n + 131072k$
		Additional energy requirements: • GPS for 12 h: 75.44 J • Accelerometer for 12 h: 927.95 J			

Now, for the aforesaid scenario, the energy requirement is estimated using Table 3.3 as follows:

(i) **Communication Energy**: Taking into account the interchange of resource inventory, query and response during node-encounter.

Transmission Energy:

Energy consumption for DPDRM:

Inventory sharing:

$180 \times 0.25 \times (32 \times 130 + 131072 \times 7) = 41474880\ \mu\text{J} = 41.5$ J (approx.)

Resource discovery:

$120 \times 0.25 \times (32 \times 130 + 9360) = 405600\ \mu\text{J} = 0.40$ J (approx.)

Energy consumption for DPD:

$300 \times 0.25 \times (24 \times 130 + 9360) = 936000\ \mu\text{J} = 0.94$ J (approx.)

Reception energy:

Energy consumption for DPDRM:

Inventory sharing:

$180 \times 0.24 \times (24 \times 130 + 131072 \times 7) = 39770956\ \mu\text{J} = 39.78$ J (approx.)

Resource discovery:

$120 \times 0.24 \times (24 \times 130 + 9360) = 359424\ \mu\text{J} = 0.36$ J (approx.)

Energy consumption for DPD:

$300 \times 0.24 \times (16 \times 130 + 9360) = 823680\ \mu J = 0.82$ J (approx.)

Total communication energy after 12 h:

DPDRM: $(41.5 + 0.40 + 39.78 + 0.36)$ J $= 82.04$ J

DPD: $(0.94 + 0.82 + 75.44 + 927.95)$ J $= 1005.15$ J

(ii) **Computation energy**:

Energy consumption for DPDRM:

Inventory sharing: $3699.2 \times 130 \times 7 + 2340.4$ pJ $= 3.37\ \mu J$ (approx.)

Resource discovery: $17124.4 \times 130 \times 3 + 4736.8 = 6.68\ \mu J$ (approx.)

Energy consumption for DPD: $1379.2 \times 130 \times 130 + 2172.6 \times 130 + 3175 = 23.59\ \mu J$ (approx.)

(iii) **Storage overhead**:

DPDRM: $8192 \times 123 \times 3 + 131072 \times 7$ bits $= 481$ KB

DPD: $16 \times 123 + 8192 \times 123 \times 3 + 131072 \times 7$ bits $= 481.24$ KB

The aforesaid illustration reveals that the communication energy requirement of the DPDRM scheme is significantly lesser than the DPD scheme. The DPD scheme typically employs GPS and accelerometers to forecast the outdoor and indoor locations of the nodes. This procedure of location prediction of the nodes demands extra energy. It is further apparent that the computational energy requirement of the DPDRM scheme is significantly lower compared to the DPD scheme. Finally, both the schemes demonstrate almost identical storage overhead.

3.3.4 *Performance Evaluation*

The DPDRM scheme is evaluated through two phases of simulation experimentations. In the first phase of the experimentation, the periodic *groupcasting* of resource inventories (*inventory sharing*) of the scheme is analyzed under different unicast and multicast routing protocols such as Epidemic, PRoPHET, MaxProp, Ge-PRoPHET, SDM and EASDM. On the other hand, in the second phase, the performance of both the *inventory sharing* and *resource discovery* of the scheme is compared with one contemporary competitor DPD [11] which works on content sharing.

3.3.4.1 Simulation Setup

The experiments are conducted using ONE simulator, version 1.5.1-RC2 [25]. The unofficial crisis map (Fig. 3.9) used during the relief operation conducted in Kathmandu of Nepal after the Nepal Earth Quake in 2015 is considered as one of the

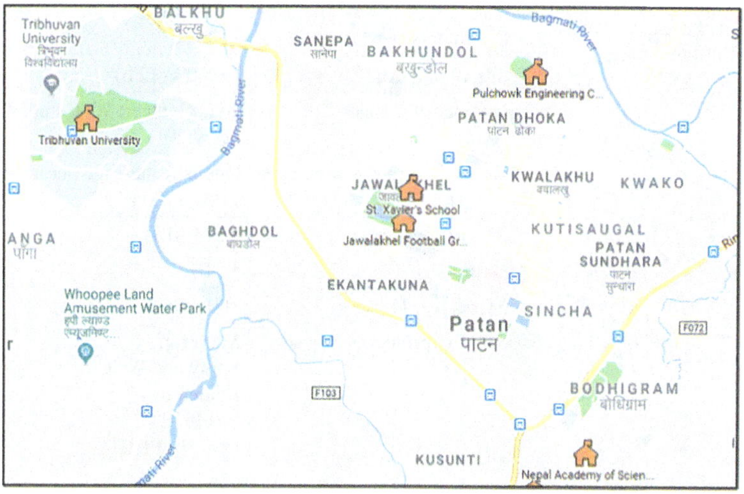

Fig. 3.9 Unofficial crisis map of Nepal earthquake [26] used for simulation

configuring parameters of ONE. Accordingly, the disaster-ravaged area of Kathmandu is considered as an experimental environment where within an area of 3.5 km^2 of Patan, four camps (*camp nodes*) are located. These camps are built in four areas, namely Pulchowk Engineering College Campus, St. Xavier's School, Jawalakhel Football Ground and Nepal Academy of Science. Also, at Tribhuvan University, the command and control center (*command and control node*) is built.

Each camp is equipped with *Throwboxes* (TBs). Volunteers equipped with smartphones are treated as *volunteer nodes*. The number of *volunteer nodes* for simulation is set to 90, based on the proposed mobile node deployment plan to be discussed in detail in Chap. 4. Two supply vehicles (*ferry nodes*) periodically visit the camps from the command and control center. It is assumed that messages are generated every 1800–2100 s. The PDM model is employed as the underlying mobility model for *volunteer nodes*. The supply vehicles follow the map route movement model. *Three-tier* architecture to be discussed in Chap. 4 is chosen for the simulation. During the simulation, *volunteer nodes* are treated as *tier-1* devices, *camp nodes* as *tier-2* devices and *ferry nodes* as *tier-3* devices.

The relief operations in a post-disaster situation usually take place in the daytime and typically it is for 12 h. Accordingly, the resource inventories get updated and shared every 12 h. Further, the resource inventories are *groupcast* periodically (once a day) from the *camp nodes*. Also, it is assumed that the queries are created randomly from the *volunteer nodes* within a simulation period of 1–2 h. Table 3.4 gives the values of other significant parameters used in the simulation.

Table 3.4 Parameter values used in the simulation

Parameters	Values
Simulation time	43200 s (12 h)
Number of nodes	96 (volunteer: 90, Camp: 4, Ferry: 2)
Generation rate of messages	One new message per node every 1800–2100 s
Size of messages	0.5–1 MB
Buffer size	5 MB
Message TTL	100 min
Routing protocols	MaxProp Spray and Wait Epidemic PRoPHET Pen-PRoPHET

3.3.4.2 Simulation Metrics

We use two kinds of metrics to assess the performance of the DPDRM. These metrics are based on the design parameter and network parameters. The design parameters control the performance of the scheme's design goals. On the other hand, the network parameters control the performance of the underlying network while achieving design goals.

In assessing *inventory sharing* during the first phase of experimentation we consider three design parameters, such as *inventory share rate*, *overhead ratio* and *average energy consumption*. Here, we consider only one parameter that is *average delivery latency* as the network parameter.

We define both the set of parameters as follows:

Inventory share rate: This metric assesses the extent to which resource inventory spreads within a fixed period of time. It is expressed as follows:

$$\text{Inventory share rate } (\%) = \frac{\text{Number of nodes received resource inventories till a particular instant of time}}{\text{Total number of nodes in the network}} \times 100\%$$

Overhead ratio: It estimates the overhead in terms of how many copies [27] of the message are passed for sending the message to its destination. The overhead ratio is formulated as follows:

$$\text{Overhead ratio} = \frac{\text{Number of messages relayed}}{\text{Total number of messages delivered}}$$

All the time the *overhead ratio* is greater than 1 for multicopy routing.

Average energy consumption: It is defined as the average energy requirement of a node for sending a message to its destination [27].

Average delivery latency: This metric measures the time required for a message to be sent [27] to its destination since its generation. It specifies how fast the message is delivered.

Further, the second phase of experimentation uses two design parameters, namely *response delivery rate* and *average energy consumption*. Similar to the first phase, it uses *average delivery latency* as the network parameter. Thus, we only need to define the *response delivery rate*.

Response delivery rate: It estimates how efficient the DPDRM is in delivering the response. The *response delivery rate* is formulated as follows:

$$\text{Response delivery rate}(\%) = \frac{\text{Number of responses delivered to the query originator nodes till a particular instant of time}}{\text{Total number of queries generated till that particular instant of time}} \times 100\%$$

3.3.4.3 Results and Discussion

In the first phase of experimentation, the performance of the proposed DPDRM scheme is evaluated only for *inventory sharing*. The DPDRM scheme uses the Ge-PRoPHET protocol for *inventory sharing* and therefore, the performance of DPDRM for *inventory sharing* is primarily dependent on the performance of such routing protocol. Thus, the said performance including results and discussion will be elaborated in Chap. 5, where the Ge-PRoPHET will be described.

The comparative performance of the DPDRM is measured in the second phase of experimentation. The comparison is primarily done with the DPD [11]. The comparison is also done with a variant of DPDRM, which eliminates the *inventory sharing* part. This variant that is DPDRM with no inventory sharing should perform better than the DPDRM w.r.t. energy overhead as the overhead incurred due to inventory sharing is removed. But the consequence of removing the sharing part on other design and network parameters is to be verified.

The comparative performance of the DPDRM is measured in terms of both the design and network parameters through three sets of experiments. Figure 3.10 shows the results based on the design parameters.

Figure 3.10a exhibits the results of the first sets of experimentation where the *response delivery rate* is measured over time. We observe that the *response delivery rate* rises progressively over time and reaches a steady-state for both the DPDRM and DPD. We also observe that the DPDRM gives the highest *response delivery rate*. For example, after 8 h (28800 s), such rates are 44%, 68% and 88% in DPDRM with no inventory sharing, DPD and DPDRM, respectively.

The primary reason behind the better *response delivery rate* demonstrated by the DPDRM is due to the availability of sufficient resource inventory provider nodes. This has been achieved by the introduction of the *inventory sharing* module which enables the dissemination of resource inventories among the *volunteer nodes* of

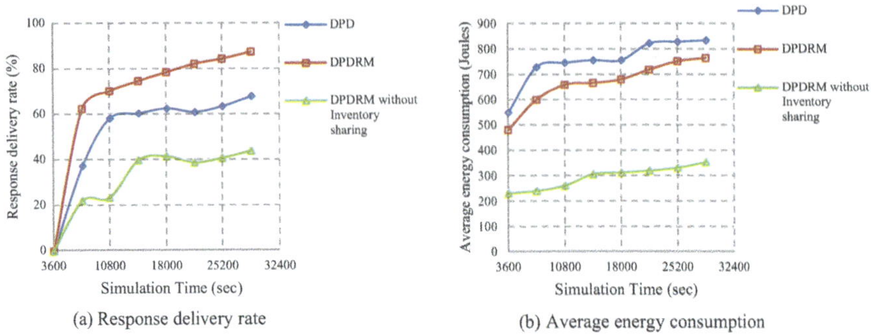

Fig. 3.10 Comparative evaluation of DPDRM with respect to design parameters

designated response groups. Hence, the *response delivery rate* improves in DPDRM. On the contrary, in DPDRM without inventory sharing, resource inventories are only available with the *camp nodes* which are scarce in the network.

The key observations from the above experimentation are the proposed DPDRM can respond against 88% (approximately) queries within 8 h of the query initiation. Further, the response delivery rate of DPDRM is around 30% higher than its competitor DPD. The results prove the DPDRM's suitability in the response-time-constrained post-disaster application.

Figure 3.10b shows the results of the second set of experimentation where *average energy consumption* is compared with time. We observe that the average energy consumption rises more or less linearly over time for both the DPDRM and DPD. We further observe that such consumption is the lowest in DPDRM. Precisely, in DPDRM *average energy consumption* is more or less 11.68% less than the DPD and DPDRM with no inventory sharing.

Figure 3.11 exhibits the comparative results of the third set of experimentation in terms of network parameter *average delivery latency*. The latency is varied with

Fig. 3.11 Comparative evaluation of DPDRM with respect to the network parameter

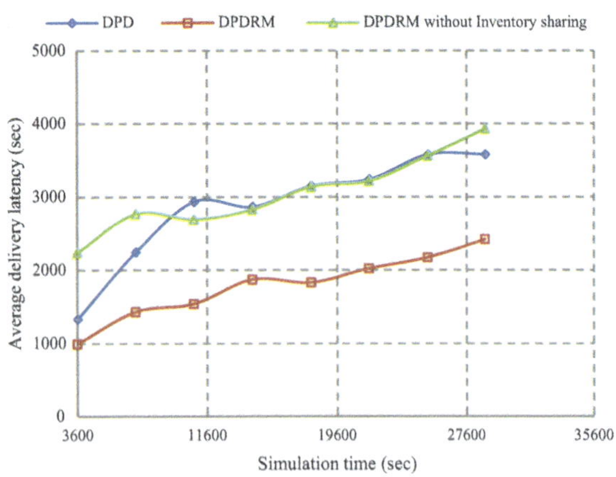

time. It is apparent from the outcomes of the experiment that the *average delivery latency* also raises more or less linearly over time for both the DPDRM and DPD. The results also show that after 8 h of simulation time, DPDRM with no inventory sharing, DPD and DPDRM show *average delivery latency* as 3934.76, 3582.78 and 2420.53 s, respectively. The results ensure the DPDRM's superiority in terms of response delivery over the DPD.

Summarily, the results of the above three sets of experiments prove that the DPDRM outperforms over DPD in terms of all the measuring parameters, such as response delivery, latency and energy consumption. Though the DPDRM with no inventory sharing shows better performance in energy efficiency, it is the worst performer with respect to the other two parameters. Thus, DPDRM with no inventory sharing is not suitable to be used in post-disaster resource management.

3.4 Conclusion

In this chapter, two post-disaster management services through Disaster Messenger and DPDRM have been discussed. The primary role of Disaster Messenger is to share situational information without the support of conventional communication infrastructure. It harnesses the power of Wi-Fi direct to offer opportunistic networks-like characteristics. The Disaster Messenger supports multithreading and hence can handle multiple send/receive operations in parallel. The experimental results disclose that Disaster Messenger outperforms SHAREit and BitTorrent Sync in terms of both the parameters such as energy consumption and data transfer rate.

On the other hand, DPDRM is proposed to automate post-disaster resource management. The scheme offers the periodic group-based distribution of resource inventories which permits effective search and retrieval of queries within the desired timeframe. The simulation results reveal the proficiency of the proposed solutions in the post-disaster scenarios in terms of *delivery ratio*, delivery latency and energy consumption. To be more specific, the DPDRM effectively responds against approximately 88% of queries consuming around 28% of the stored energy.

In order to offer the services mentioned in Chap. 2 and in this chapter, the existence of a post-disaster opportunistic network framework becomes crucial. Such a framework comprises two entities such as opportunistic networks architecture and energy-efficient post-disaster routing protocols. Thus, in the next two chapters designing a post-disaster opportunistic network framework is elaborated.

References

1. Huang, Q., Xiao, Y., Huang, Q., Xiao, Y.: Geographic situational awareness: mining tweets for disaster preparedness, emergency response, impact, and recovery. ISPRS Int. J. Geo-Inf. **4**, 1549–1568 (2015). https://doi.org/10.3390/ijgi4031549

2. Madey, G.R., Barabási, A.-L., Chawla, N.V., Gonzalez, M., Hachen, D., Lantz, B., Pawling, A., Schoenharl, T., Szabó, G., Wang, P., Yan, P.: Enhanced Situational Awareness: Application of DDDAS Concepts to Emergency and Disaster Management (2007)
3. Ramchurn, S.D., Huynh, T.D., Ikuno, Y., Flann, J., Wu, F., Moreau, L., Jennings, N.R., Fischer, J.E., Jiang, W., Rodden, T., Simpson, E., Reece, S., Roberts, S.J.: HAC-ER: A disaster response system based on human-agent collectives. In: Proceedings of the 2015 International Conference on Autonomous Agents and Multiagent Systems, vol. 1, pp. 533–541 (2015). https://doi.org/10.1613/jair.5098
4. Sardouk, A., Mansouri, M., Merghem-Boulahia, L., Gaiti, D., Rahim-Amoud, R.: Multi-agent system based wireless sensor network for crisis management. In: 2010 IEEE Global Telecommunications Conference GLOBECOM 2010, pp. 1–6. IEEE (2010)
5. Luo, C., Nightingale, J., Asemota, E., Grecos, C.: A UAV-cloud system for disaster sensing applications. In: 2015 IEEE 81st Vehicular Technology Conference (VTC Spring), pp. 1–5. IEEE (2015)
6. Landwehr, P.M., Carley, K.M.: Social Media in Disaster Relief (2014)
7. Vieweg, S., Hughes, A.L., Starbird, K., Palen, L.: Microblogging during two natural hazards events. In: Proceedings of the 28th International Conference on Human Factors in Computing Systems—CHI '10, p. 1079. ACM Press, New York, NY, USA (2010)
8. Yin, J., Lampert, A., Cameron, M., Robinson, B., Power, R.: Using social media to enhance emergency situation awareness. IEEE Intell. Syst. **27**, 52–59 (2012). https://doi.org/10.1109/MIS.2012.6
9. Sen, A., Rudra, K., Ghosh, S.: Extracting situational awareness from microblogs during disaster events. In: 2015 7th International Conference on Communication Systems and Networks (COMSNETS), pp. 1–6. IEEE (2015)
10. Pitkanen, M., Karkkainen, T., Greifenberg, J., Ott, J.: Searching for content in mobile DTNs. In: 2009 IEEE International Conference on Pervasive Computing and Communications, pp. 1–10. IEEE (2009)
11. Talipov, E., Chon, Y., Cha, H.: Content sharing over smartphone-based delay-tolerant networks. IEEE Trans. Mob. Comput. **12**, 581–595 (2013). https://doi.org/10.1109/TMC.2012.21
12. Chen, K., Shen, H., Yan, L.: Efficient file search in delay tolerant networks with social content and contact awareness. IEEE Trans. Parallel Distrib. Syst. **27**, 1982–1995 (2016). https://doi.org/10.1109/TPDS.2015.2472005
13. Hyytiä, E., Bayhan, S., Ott, J., Kangasharju, J.: On search and content availability in opportunistic networks. Comput. Commun. **73**, 118–131 (2016). https://doi.org/10.1016/j.comcom.2015.09.011
14. Kaisar, S., Kamruzzaman, J., Karmakar, G., Gondal, I.: Content sharing among visitors with irregular movement patterns in visiting hotspots. In: Proceedings of the 2015 IEEE 14th International Symposium on Network Computing and Applications, NCA 2015, pp. 230–234 (2016). https://doi.org/10.1109/NCA.2015.16
15. SHAREit-Discover and Share Unlimited Joy. https://www.ushareit.com/en/index.html
16. BitTorrent Sync: Faster file synchronization. https://www.resilio.com/
17. Bhattacharjee, S., Kanta, S., Modi, S., Paul, M., Dasbit, S.: Disaster messenger: an android based infrastructure less application for post disaster information exchange. In: 2016 IEEE International Conference on Advanced Networks and Telecommunications Systems, ANTS 2016 (2017)
18. Zhang, L., Tiwana, B., Qian, Z., Wang, Z., Dick, R.P., Mao, Z.M., Yang, L.: Accurate online power estimation and automatic battery behavior based power model generation for smartphones. In: Proceedings of the Eighth IEEE/ACM/IFIP International Conference on Hardware/Software Codesign and System Synthesis—CODES/ISSS '10, p. 105. ACM Press, New York, NY, USA (2010)
19. Bhattacharjee, S., Roy, S., DasBit, S.: DPDRM: a decentralized post-disaster resource management scheme using energy efficient smart phone based DTN. J. Netw. Comput. Appl. **111** (2018). https://doi.org/10.1016/j.jnca.2018.03.007

20. Cao, Y., Sun, Z.: Routing in delay/disruption tolerant networks: a taxonomy, survey and challenges. IEEE Commun. Surv. Tutor. **15**, 654–677 (2013). https://doi.org/10.1109/SURV.2012.042512.00053
21. Uddin, Y.S., Nicol, D.M., Abdelzaher, T.F., Kravets, R.H.: A post-disaster mobility model for delay tolerant networking. In: Proceedings of the Winter Simulation Conference, pp. 2785–2796 (2009). https://doi.org/10.1109/WSC.2009.5429249
22. Perrucci, G.P., Fitzek, F.H.P., Widmer, J.: Survey on energy consumption entities on the smartphone platform. In: 2011 IEEE 73rd Vehicular Technology Conference (VTC Spring), pp. 1–6. IEEE (2011)
23. Kavvadias, N., Neofotistos, P., Nikolaidis, S., Kosmatopoulos, C., Laopoulos, T.: Measurements analysis of the software-related power consumption in microprocessors. IEEE Trans. Instrum. Meas. **53**, 1106–1112 (2004). https://doi.org/10.1109/TIM.2004.830784
24. Wang, W., Zwolinski, M.: An improved instruction-level power model for ARM11 microprocessor (2014)
25. Bhattacharjee, S., Chatterjee, R., Pal, T., DasBit, S.: Implementing multicasting and broadcasting of multimedia data in ONE simulator. In: Proceedings of the 10th EAI International Conference on Simulation Tools and Techniques—SIMUTOOLS '17, pp. 48–56 (2017). https://doi.org/10.1145/3173519.3173540
26. Nepal Quake Unofficial Crisis Map—Google My Maps. https://www.google.com/maps/d/viewer?mid=1_PvxC6bAC9zpk86jRHvAuLCPJ3k&ll=27.692258449090865%2C85.33920080859377&z=10
27. Bhattacharjee, S., Bit, S. Das: EnTER: an encounter based throwbox deployment strategy for enhancing network reliability in post-disaster scenarios over DTN. In: Proceedings of the 20th International Conference on Distributed Computing and Networking—ICDCN '19, pp. 413–416. ACM Press, New York, NY, USA (2019)

Chapter 4
Reliable and Energy-Efficient Post-disaster Opportunistic Network Architecture

In order to offer the post-disaster management services discussed in the last two chapters, the existence of a post-disaster opportunistic network framework becomes crucial. A post-disaster opportunistic network framework comprises two entities, such as opportunistic network architecture and energy-efficient post-disaster routing protocols. The post-disaster opportunistic network architecture typically includes suitable network architecture and node deployment plans which are elaborated in this chapter.

Disasters usually destroy the conventional communication infrastructure relentlessly causing challenges to the disaster response agencies while communicating among themselves in the disaster aftermath. In addition to that, a well-organized disaster response operation necessitates the sharing of situational information and resource inventories effectively between different disaster response agencies. Moreover, map-based navigation assistance for the volunteers and victims is also crucial in disaster response operations. Due to the intermittent network connectivity and unstable power supply, it is challenging to establish a fully connected provisional network supporting such services in post-disaster scenarios. The research fraternity has encouraged the use of opportunistic networks as a feasible substitute for exchanging information during post-disaster scenarios [1–3]. The physical mobility of the volunteers, carrying mobile handheld devices such as smartphones, tablets, etc. are typically exploited by opportunistic networks as communication opportunities. The communication among the mobile handheld devices is established when two such devices arrive within the radio range of each other. This kind of network may be deployed in the affected areas to offer elementary communication capability.

In order to implement opportunistic networks effectively in post-disaster scenarios, designing a suitable architecture is of utmost importance. The post-disaster opportunistic network architecture typically includes suitable network architecture and node deployment plans. During the design of an opportunistic network architecture for post-disaster communication, there is a couple of important parameters, viz., (i) network reliability and (ii) energy efficiency that are to be taken care of.

S. Bhattacharjee et al., *Post-disaster Navigation and Allied Services over Opportunistic Networks*, Smart Innovation, Systems and Technologies 228,
https://doi.org/10.1007/978-981-16-1240-4_4

In post-disaster scenarios, network reliability is expressed in terms of enhanced message delivery with reduced latency. Due to the unavailability of the end-to-end path between the source and the destinations, the exchange of messages in opportunistic networks depends merely on contact opportunity (encounters) between the nodes, which may be static or mobile. An increasing number of encounters among the nodes leads to enhanced message delivery with reduced latency, which in turn improves network reliability. The contact opportunity among the nodes typically depends on the node deployment. Moreover, most of the devices, which are used in opportunistic networks, are primarily battery-powered and hence get depleted over time. Thus, these types of networks have a limited lifetime after which devices need to be re-energized. So, an opportunistic network-based communication system should have energy-efficient network architecture in order to provide reliable network connectivity for a longer period.

Although extensive research has been conducted to develop effective network architecture and suitable node deployment plans for opportunistic networks, very limited of them are intended for post-disaster communication [4–9]. To the best of our knowledge, the cost of installation and the performance of such opportunistic network architectures under different opportunistic network routing protocols in terms of energy consumption, data delivery and overhead have not been studied. In addition to that, the aforesaid architectures have not been tested in post-disaster scenarios under the post-disaster mobility model. Hence, existing opportunistic network architectures prove unsuitable for post-disaster communication. Thus, in this chapter, we design [10] an energy-efficient network architecture with a suitable node deployment plan for reliable post-disaster communication.

4.1 Literature Review

The works on the opportunistic network, including architecture, node deployment, routing and mobility models are reviewed here.

4.1.1 Architecture

It had been observed that extensive studies have been carried out to develop network architecture ensuring maximum message delivery with minimum energy consumption in opportunistic networks. Diverse architectures starting from flat to multitier [4–9, 11–13] are also designed for opportunistic networks considering different application necessities. Yet, only a limited quantity of those aforesaid architectures [4–9] have been studied from the post-disaster communication viewpoint. All opportunistic network architectures proposed till date can be roughly categorized into three such as two-tier, three-tier and four-tier architecture. The following sub-sections review the literature on different categories of opportunistic network architecture.

4.1.1.1 Two-Tier Architecture

Two-tier architecture typically consists of mobile nodes in *tier-1* and static relay nodes in *tier-2*. Some of the works on two-tier opportunistic network architecture are discussed here.

In one of the works [11], an opportunistic network architecture is proposed where energy-efficient, low-cost, battery-powered, static nodes or *Throwboxes* (TBs) are employed to improve the message delivery. The introduction of TBs enhances network throughput and reduces delivery latency. However, the degradation of performance is evident in such network architecture when the nodes follow an unstructured (random) mobility pattern. The volunteers typically exhibit event-driven mobility pattern (neither purely random nor periodic) in post-disaster scenarios [14], which is not a favorable condition for implementing this architecture.

In another work [12], a *two-tier* opportunistic network architecture consisting of static nodes and autonomous flight drones (AFWs) is proposed. The AFWs facilitate the transmission of situational information from the residences or evacuation shelters to the isolated disaster management headquarter by flying between them periodically. This architecture proves suitable for exchanging situational information between isolated establishments. However, evaluation of energy consumption which is a crucial requirement for the implementation of opportunistic network architecture in post-disaster scenarios is not performed in [12].

4.1.1.2 Three-Tier Architecture

In addition to the mobile and static nodes employed in *two-tier* architecture, the *three-tier* architecture employs an additional type of node for message routing. Such nodes may be static or mobile depending on the application scenarios.

In one such work, an architecture called *DakNet* [13] is proposed which exploits existing network infrastructure and transportation infrastructure to offer communication facilities to the villages. *DakNet* consists of (i) kiosks, (ii) mobile access points (MAP) and (iii) Hubs (internet access points). It is a low-cost architecture that provides network communication in challenged areas. Here, kiosks are typically placed in remote areas and hubs are placed in cities. However, in order to function properly, the kiosks need a steady power supply which might be ruled out in affected areas.

In another work [4], a rapidly deployable *three-tier hybrid wireless mesh network* (HWMN) architecture is proposed as a part of the Responding to Crises and Unexpected Events (RESCUE) project. Here, the first tier consists of mobile devices like PDA, laptops and iTAGs carried by the volunteers. Second tier consists of wireless access points, and the gateways typically form the third tier.

In [5], another network architecture (*DistressNet*) is presented for the post-disaster scenario. The *DistressNet* architecture consists of several sub-networks, such as (i) *BodyNet*, (ii) *TeamNet*, (iii) *Vehicle Net*, (iv) *AreaNet* and (v) *SenseNet*. This architecture is primarily intended for the affected areas and can offer energy-efficient data communication services to the volunteers.

In another work [6], *AnonymousNet (AnonNet)* is proposed which offers a low-cost communication alternative for large affected areas. *AnonNet* comprises of second-generation wireless sensors, ad hoc and opportunistic networks for disaster response. *AnonNet* typically consists of four categories of devices, such as (i) monitoring/sensing devices, (ii) end-user interactive devices, (iii) network backbone and (iv) fixed mesh. Although this system provides a low-cost communication solution for the affected areas, the technical expertise required to maintain this system reduces its viability significantly where adequate technical support is unavailable.

4.1.1.3 Four-Tier Architecture

The *four-tier* architecture is primarily intended for such affected areas that are inaccessible due to disaster-related destructions. This architecture employs additional static nodes with long-range communication capability along with the three types of nodes employed in the *three-tier* architecture.

In [7, 8] a *four-tier* hybrid architecture is proposed for post-disaster communication by employing smartphones, *Dropboxes*, *data mules* and non-line-of-sight (NLOS) Wi-Fi towers. The Wi-Fi towers are primarily used to offer high throughput and low latency data communication to the different shelters of the affected area. On the contrary, the utilization of the NLOS Wi-Fi towers in this architecture increases overhead in terms of energy consumption and cost of deployment. This architecture is suitable for such affected areas which are isolated due to the severity of the disaster.

In another work [9], a four-tier architecture for post-disaster situation analysis is presented. Here, the entire affected area is considered as the collection shelter points (SP). This multitier architecture consists of mobile Relief Workers (RW) at *tier-1*, and *Throwboxes* (TBs) at *tier-2*. The vehicles, boats, etc. that provide inter-SP connectivity are employed as *Data mules* (DM) at *tier-3*. The master control station (MCS) which is situated at a remote location from the affected area at *tier-4*. In this architecture, the information from the affected areas is accumulated by the RW equipped with smartphones and forwarded to the TBs. The information stored at the different TBs is forwarded to the MCS with the help of the DMs. The said architecture facilitates reliable data dissemination through opportunistic networks.

The majority of the aforesaid opportunistic network architectures possess some limitations which restrain them to be deployed in the affected areas in post-disaster scenarios. One of the predominant limitations is the lack of estimation of energy efficiency. It has been observed that the energy consumption of network architecture has been evaluated only in [6, 11]. To the best of our knowledge, the cost of installation and the performance of such opportunistic network architectures under different opportunistic network routing protocols in terms of energy efficiency, data delivery and overhead have not been evaluated. Hence, none of the aforesaid opportunistic network architectures proves suitable for post-disaster communication.

4.1.2 *Node Deployment*

Node deployment is a crucial aspect of opportunistic network architecture. It has been observed that message forwarding in opportunistic networks is accomplished by employing the store–carry–forward paradigm, where node mobility creates communication opportunities. Hence, the performance of opportunistic network architecture with respect to message delivery and delay depends on the deployment of nodes (network topology) which is highly dynamic. Opportunistic network architecture consists of heterogeneous nodes (e.g. smartphones, tablets, laptops, *Throwboxes* and different sensor nodes). Such nodes are broadly categorized into two, such as mobile nodes and static nodes. The literature review regarding mobile and static node deployment is elaborated below.

4.1.2.1 Mobile Node

The message dissemination over opportunistic networks is dependent on the encounter among the mobile and static nodes. The mobile nodes carry messages to the destinations in single or multiple hops through their physical mobility. Hence, in order to design reliable opportunistic network architecture for post-disaster scenarios, an appropriate approach for deploying mobile nodes becomes indispensable. Formulating a suitable strategy for deploying mobile nodes becomes challenging due to the frequent change in the position of the nodes. Hence, in the context of opportunistic networks, mobile node deployment refers to the number of nodes present in a unit area (node density). A very few works are reported which explore the mobile node deployment in opportunistic networks. Such works are mentioned below.

In one such work [15], a graph-based model for deploying mobile node deployment in an opportunistic network is proposed. The authors prove that the connectivity patterns of mobile nodes over time can be modeled as Erdos-Renyi random graphs, but this approach assumes a specific class of mobility model called Marks-Memoryless which is unsuitable for post-disaster scenarios.

In another work [16], the authors analyze the extent of the message delivery in time-varying networks such as opportunistic networks. They model the encounter among the nodes in such a network as Markovian random graphs. They study the message forwarding in aggregated-Markovian random graphs and derive that the extent of message delivery is dependent on the node density in the network. Although this work models the connectivity pattern of nodes as an edge-Markovian graph, Pebeyre et al. [17] reveal that the Markovian graph model assumes a uniform distribution of mobile nodes under realistic mobility traces which is infeasible in post-disaster scenarios.

Thus, the aforesaid limitations restrict the existing works to be implemented as a mechanism for mobile node deployment in post-disaster scenarios. Apart from opportunistic networks, a substantial amount of work has been conducted on mobile node deployment in related fields such as WSN. Some of the works in [18, 19] are relatively recent reporting. The primary aim of those works is to enhance network coverage and lifetime by employing self-deployable mobile sensor nodes. However, unlike opportunistic networks, the mobility of sensor nodes in WSN is restricted within a tiny area for a limited period. Hence, the techniques for deploying mobile sensor nodes in WSN do not prove suitable for opportunistic networks.

4.1.2.2 Static Node

During post-disaster scenarios, the nodes demonstrate uncertain encounters among themselves due to the non-uniformity in the movement patterns. Such encounters result in undesirable message drop and poor delivery of messages. Thus, the overall network reliability is downgraded.

In an opportunistic network, the employment of a restricted quantity of static nodes (*Throwboxes*) has been an extensively adopted mechanism to enhance network reliability. The tiny low-cost battery-powered equipment fitted with wireless interfaces and large secondary storage is typically used as *Throwboxes* (TBs). Numerous approaches for deploying TBs have been suggested in the opportunistic networks literature. Such TB deployment approaches can be largely categorized into two, viz., random placement and planned placement. In random placement approaches, the TBs are typically placed in arbitrary positions irrespective of investigating the network parameters. Qirtas et al. [20] identify that although in random placement approaches the cost of deploying TBs is negligible, such approaches seldom enhance network reliability in terms of message delivery. Hence, the random placement approaches remain inappropriate in post-disaster scenarios.

In the planned placement approach the TBs are placed in a strategic manner considering several factors comprising social and network parameters. Such strategies [21–25] can be further sub-divided into two categories, viz., hotspot-based placement [21, 22] and strategic placement [23–25].

In hotspot-based placement TBs are deployed in popular locations identified by the mobility pattern of nodes. A couple of recently reported hotspot-based placement strategies are reviewed here.

In one such work [21] different frequently visited location-based (e.g. community home) TB deployment strategy is proposed. The authors assume that each community home is equipped with a TB. The scheme also introduces the notion of "virtual Throwbox" in the absence of real TB. The nodes which often visit a particular place or those that are at home are treated as "virtual Throwbox" for that particular home. This technique introduces the fault-tolerant capability in opportunistic networks.

In another work [22], a movement pattern-aware TB placement strategy is proposed in which the nodes use the working day movement model. In this scheme, the TBs are typically placed at offices, homes and other meeting spots in order to enhance the performance of the network.

Summarily, the hotspot-based placement strategy performs well under predictable mobility pattern of nodes. However, the mobility pattern demonstrated by the nodes in disaster aftermath is non-uniform as well as event-driven and thus unpredictable in nature. Therefore, a hotspot-based placement strategy does not suit this scenario.

In strategic placement, the deployment locations of TBs are identified based on several factors, like the scale of the network, topology and movement pattern of nodes, etc. Some of the well-known strategic placement schemes are discussed here.

In one such work [23], a couple of greedy algorithms are proposed for the placement of k number of TBs in opportunistic networks. One algorithm is *GrdAddTBs* in which TBs are added gradually in a network until the k number of TBs are placed. The other algorithm is *GrdDelTBs* in which initially maximum numbers of TBs are placed in the network. Subsequently, the TBs are removed from the network gradually until k TBs are left. The strategy is dependent on periodic addition or removal of TBs to reach the optimal solution.

In another work [24], several criteria are identified for deploying a fixed quantity of TBs depending on the social behavior of mobile nodes. The authors propose five different TB placement strategies to enhance network reliability. One of them places TBs in the most popular locations. However, for deploying a very large number of TBs the proposed strategy exhibits almost identical results as random placement.

In another work [25], the notion of latency graph is introduced to predict the communication latency of opportunistic networks. The authors suggest an empirical relay placement scheme for latency minimization (RPLM) using the latency graph. The proposed scheme works in two steps. During the initial step, the latency graph is divided into numerous groups using Gonzalez algorithm. In the final step, the shortest path between the center nodes of such groups is determined. The relays are placed on such shortest paths. In this scheme, the latency is reduced at the cost of overhead caused by frequent grouping in the latency graph.

In the case of strategic placement, prior knowledge regarding the network (e.g. the scale of the network, topology and movement pattern of nodes, etc.) is indispensable. Typically in scenarios such as post-disaster scenarios, the accessibility of past information regarding the network is non-existent. Moreover, the strategy proposed in [23] is dependent on the periodic addition or removal of TBs to reach the optimal solution. This strategy proves impractical in post-disaster scenarios. Hence, the existing static node deployment strategies prove unsuitable for post-disaster scenarios.

4.2 Network Architecture for Post-disaster Communication

In this section, the impact of well-known mobility models (e.g. *map-based movement model*, *random waypoint model*, *cluster movement model* and PDM model), as well as widely accepted opportunistic networks routing protocols, is investigated on different architectures of opportunistic networks to find out a suitable architecture for post-disaster communication. In order to perform the aforesaid analysis, the existing architectures of opportunistic networks have been categorized based on the types of devices present in the network at diverse hierarchical levels (tiers), their roles and functionality. Subsequently, we try to design an efficient network architecture that can support the basic situational awareness, query-response and map-based navigation services that are required to perform a post-disaster response operation. Finally, the performance of different network architecture categories is compared with the proposed architecture with respect to *delivery ratio*, energy efficiency, average latency and *overhead ratio* by applying different opportunistic networks routing protocols and mobility models.

4.2.1 Classification of Existing Network Architectures

It has been observed that numerous varieties of devices are employed in diverse architectures of opportunistic networks to facilitate different operations. Such devices demonstrate interesting resemblance in terms of operations in various layers of network architecture. Nearly all the architectures of opportunistic networks typically include three extensive categories of devices with respect to their operations.

The devices that belong to the first category typically include a group of static or mobile handheld devices (e.g. smartphones, tablets) and sensor nodes having restricted storage capability. The primary functionality of these devices is to gather situational information from the affected areas and forward them to the static devices with large storage capacity or other mobile devices in the vicinity. Such devices facilitate data dissemination in the adjacent areas by exploiting their mobility and encounter with other devices.

The devices that belong to the second category are generally stationary with large storage and battery capacity. These devices are usually deployed in buildings, road crossings or vehicles. *Throwboxes* (TB), *Dropboxes*, *wireless access points*, *mobile access points (MAP)*, etc. comes under this device category. These devices collate situational information gathered by the first category devices and forward them to the nearby devices upon encounter. The buffer overflow problem which causes unwanted message drops in opportunistic networks is typically eliminated with the introduction of second category devices. These devices also work as data forwarder and create additional contact opportunities among static and mobile devices.

The devices belonging to the third category comprise static or mobile devices which facilitate message routing in opportunistic networks. Devices such as *data mules*, *gateways*, *hubs* (Internet access points), etc. come under this category. Such devices facilitate routing either by using their physical mobility or through an additional communication interface. For instance, routing through *data mules* is provided by exploiting mobility, whereas *gateways*, *hubs*, etc. facilitate routing through traditional internet connectivity.

Another rarely used device in opportunistic network architecture is the non-line-of-sight *Wi-Fi towers*. Such devices offer long-range communication and are usually placed in the areas disconnected due to the disasters. The aforesaid observations indicate the existence of four categories of devices (*tier-1*, *tier-2*, *tier-3* and *tier-4*) in opportunistic network architecture based on the hierarchy of operation. Table 4.1 summarizes such device categorization.

Table 4.1 Categorization of communication devices based on the hierarchy of functionality

Category	Devices
Tier-1	**Mobile devices**: PDAs [4], laptops [4], smartphones [6, 7], tablet PCs [6], BTag [6], body sensors [5, 6], iTAGs [4] **Static devices**: kiosks [13], seismic sensors [6]
Tier-2	**Mobile devices**: MAPs [13] **Static devices**: Throwboxes [11], Dropboxes [7], data waypoints [6], wireless access points [4]
Tier-3	**Mobile devices**: Data mules [7] **Static devices**: Mesh router [6], extenders [6], hubs [13], gateways [4]
Tier-4	NLOS Wi-Fi towers [7]

Table 4.2 Categorization of existing architectures of opportunistic networks

Category	Existing opportunistic networks architectures
Two-tier architecture	Implementation of Throwbox [11]
Three-tier architecture	AnonNet [6] DistressNet [5] DakNet [13] HWMN [4]
Four-tier architecture	Four-tier hybrid architecture [7]

Moreover, the existing architectures of opportunistic networks are also categorized in Table 4.2. Based on the aforesaid classifications, accessible devices in post-disaster scenarios for different hierarchical levels (tiers) of network architecture are identified. Subsequently, the design of suitable opportunistic network architecture for post-disaster communication is elaborated. The goals of the opportunistic network architecture intended for post-disaster communication are illustrated as follows.

4.2.2 Design Goals

A reliable communication system plays an indispensable role in the modern post-disaster response operation. Hence, network reliability should be considered as an important parameter while designing opportunistic network architecture for post-disaster communication. In opportunistic networks, messages are disseminated using the likelihood of contact (encounter) among the devices carried by human beings or motorized and non-motorized vehicles, in the absence of end-to-end paths between the source and the destinations. The movement patterns of human beings and vehicles are mostly event-driven in post-disaster scenarios and hence unpredictable in nature. As a result, predicting contact opportunities among mobile devices become challenging in such scenarios. Therefore, uncertain message delivery and unpredictable delays in delivering messages are the inherent features of such a network.

Thus, improved message delivery with reduced delay is the primary requirement for network reliability and that is one of the design goals of network architecture intended for post-disaster scenarios.

Further, most of the devices (nodes) which are proposed to be used in opportunistic network in such scenarios run primarily on batteries, hence need to be recharged after some time. The lifespan of any opportunistic networks relies on the lifespan of the nodes which is further dependent on their residual energy. So, opportunistic network architecture for post-disaster communication should be energy-efficient such that reliable communication can be provided for a longer period of time. Thus, energy efficiency becomes another design goal for post-disaster network architecture.

In addition to this, the energy consumption of devices in an opportunistic network is proportionate to the number of transmissions initiated by the devices. Increasing replication of messages increases transmissions, which in turn increases the network congestion; hence, the delivery cost of the messages is increased. The reduced delivery cost of messages in terms of replication is another important design goal for such network architecture. The parameters influencing the design of opportunistic network architecture for post-disaster communication are elaborated in the next section.

4.2.3 Design Parameters

Based on the design goals discussed in Sect. 3.1.2, network reliability, energy efficiency and delivery cost of messages in terms of replication are identified as three crucial parameters for designing opportunistic networks-based post-disaster network architecture. The illustration of such design parameters is as follows:

- **Network reliability**: This parameter indicates the efficiency of the network architecture in the successful and timely delivery of messages to the destinations. Hence, network reliability in opportunistic networks is measured using a couple of metrics such as *delivery ratio* and *average delivery latency*. The *delivery ratio* determines the efficiency of delivering messages in opportunistic networks architecture. The *average delivery latency* determines the amount of time required from the creation of a message to its delivery. This metric specifies how fast a message can be disseminated in the network.
- **Energy efficiency**: This parameter assesses the lifetime of devices employed in the different hierarchical levels of opportunistic networks-based post-disaster network architecture. The energy efficiency is measured by the metric *average energy consumption*. This metric is calculated using the arithmetic mean of energy consumption for delivering a single message within a fixed time duration.
- **Delivery cost of messages in terms of replication**: The cost of delivering a message to its destination in terms of network congestion is estimated by the metric *overhead ratio*. The *overhead ratio* identifies the delivery cost of a message with respect to the number of message replications.

Depending on the aforesaid parameters, the design of a suitable opportunistic network architecture for post-disaster communication is illustrated in the next section.

4.2.4 Proposed Three-Tier Network Architecture

In this section, the design of the *three-tier* network architecture is illustrated for post-disaster communication. In a post-disaster scenario different disaster response agencies establish camps at suitable locations nearby the affected areas. A group of volunteers equipped with smartphones (*volunteer nodes*) is associated with each camp to accomplish different operations. The operations typically include rescuing the victims, assessing the damages by exploring the affected areas, distributing relief materials at shelters and estimating resource requirements. The *volunteer nodes* are typically mobile and demonstrate low movement speed. These nodes are primarily employed for accumulating information from the adjacent parts of the affected areas. Here, *volunteer nodes* construct the *tier-1* of the network architecture. Due to the insufficient number of volunteers and their non-uniform mobility pattern in post-disaster scenarios, intra-tier contact opportunity among *tier-1* devices becomes limited.

Smartphones or tablets with large storage and battery capacity positioned in the camps or road crossings are preferred as *Throwboxes* (*tier-2* devices) in the proposed architecture. Camps equipped with *Throwboxes* are considered as *camp nodes*. *Throwboxes* are typically static and employed as persistent storage for situational information collected by *volunteer nodes*. The introduction of *Throwboxes* creates additional inter-tier contact opportunities among *tier-1* and *tier-2* devices.

Supply vehicles equipped with smartphones or tablets having extended storage are chosen as *ferry nodes* (*tier-3* devices) in the proposed *three-tier* network architecture. Such vehicles periodically visit different camps within a region from the command and control center to supply resources. It has been observed that an entire region is not affected by the disaster consistently. Depending on the geography some parts are brutally affected. Thus, the affected areas are usually distributed inside a region. Collecting situational information from such remote disaster-affected areas through *tier-1* devices periodically becomes infeasible due to their low movement speed. Hence, *ferry nodes* are employed in order to collect situational information from such areas within a short duration. *Ferry nodes* are primarily mobile and offer message routing among remote-affected areas through their hi-speed physical mobility.

Figure 4.1 illustrates a snapshot of the proposed *three-tier* opportunistic network architecture consisting of four *camp nodes* (*tier-2*) and eight shelters in two affected areas in a region. Each area comprises five *volunteer nodes* (*tier-1*). The command and control center is set up outside such areas. A supply vehicle (*ferry node*) periodically visits each *camp node* from the command and control center. The *volunteer nodes* visit the shelters from the *camp nodes* in order to distribute relief. During such visits, they accumulate situational or resource requirement information and

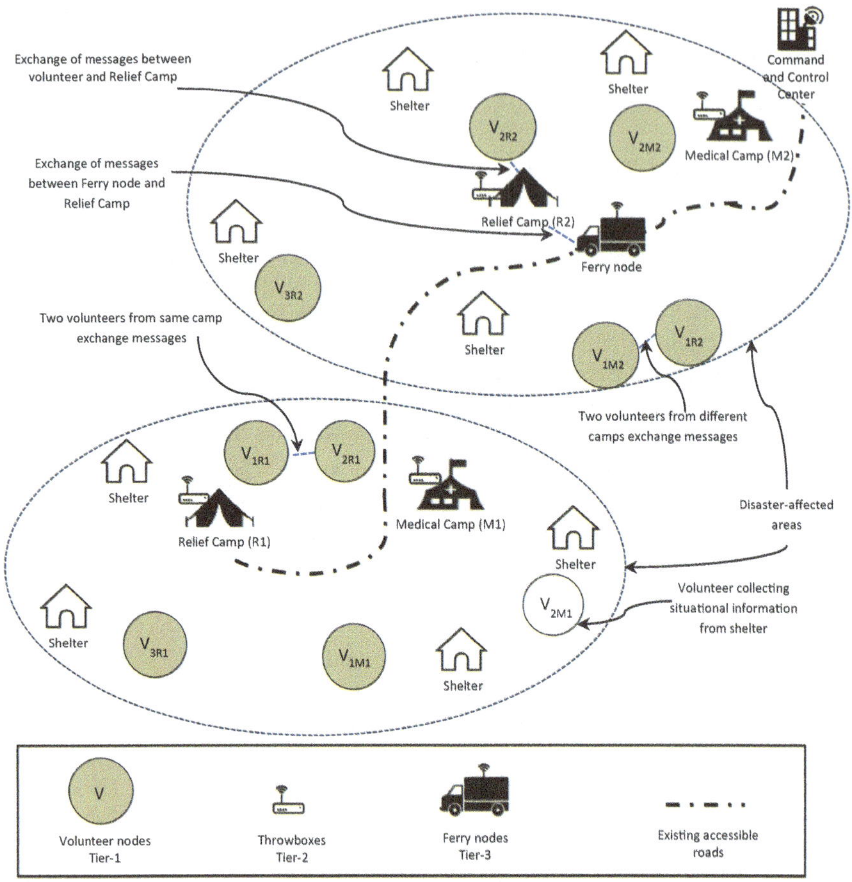

Fig. 4.1 Proposed *three-tier* opportunistic network architecture for post-disaster communication

subsequently forward such information to the *camp nodes*. The *camp nodes* collate such information acquired from the different parts of the affected areas and share the collated information with other *camp nodes* and the command and control center through the *ferry node* (*tier-3*).

The command and control center acquires such information from the camps situated in different parts of the affected areas and constructs a unified global view of the post-disaster situation. Such global views of post-disaster scenarios assist the response agencies in decision making.

4.2.5 *Performance Evaluation*

The performance of the proposed *three-tier* architecture is assessed through simulation in ONE simulator, version 1.5.1-RC2 [26]. The evaluation is done by comparing it with the existing *two-tier* [11] and *four-tier* [7] architectures in terms of design parameters using different well-known routing protocols and mobility models of opportunistic networks. The simulation setup is constructed depending on the post-disaster response operation carried out after the flash flood that occurred in the Northern Indian state, Uttarakhand in June, 2013.

4.2.5.1 Simulation Setup

The simulation experiments are conducted using ONE simulator, version 1.5.1-RC2. The impact map availed by Google crisis response (Fig. 4.2) used during rescue/relief operation conducted at Badrinath and Rudraprayag of Uttarakhand after the flash flood in 2013 is considered as one of the configuring parameters of ONE.

Initially, the simulation environment is set up for *two-tier* architecture. In the disaster-ravaged area of Uttarakhand, five relief camps of five disaster-response agencies are located. Volunteers equipped with smartphones are treated as *volunteer nodes*. The number of *volunteer nodes* (*tier-1*) for simulation is set to 105. They belong to the said relief camps. Each of the relief camps is equipped with *Throwboxes* (TBs) which is considered as *tier-2*. The nodes in both the *tier*-1and *tier-2* are randomly deployed. The relief operations in a post-disaster scenario are usually

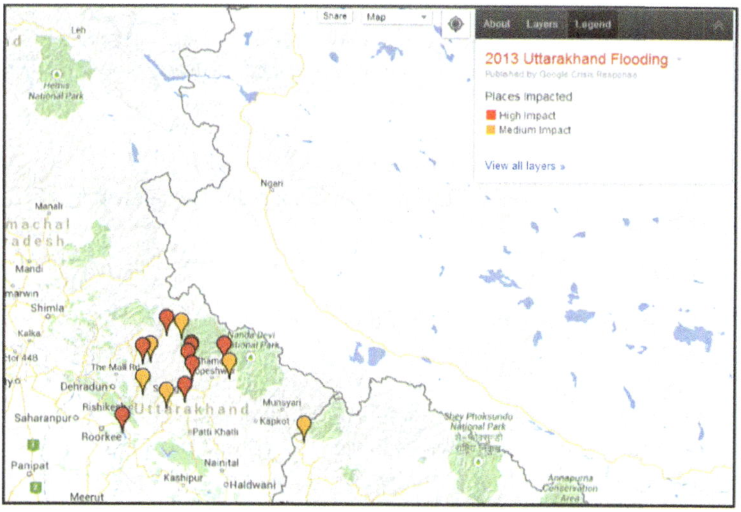

Fig. 4.2 Disaster impact map of Uttarakhand, India [27]

conducted in the daytime and typically it is for 12 h. Accordingly, the resource inventories get updated and shared every 12 h (43200 s) which is chosen as simulation time.

Next, the simulation environment is set up for *three-tier* architecture by introducing a concept of *ferry nodes*. Such *ferry nodes* are considered to work at *tier-3*. Supply vehicles considered as *ferry nodes* periodically visit the camps from the command and control center. The supply vehicles are equipped with smartphones having additional storage. These nodes follow uniform mobility patterns and move faster than the nodes working at *tier-1* and *tier-2*.

Finally, the simulation environment is set up for *four-tier* architecture by augmenting *three-tier* architecture with additional high-speed, long-range devices having a radio transmission range around 2 km at *tier-4*. These devices are relatively high-end compared to the devices at other *tiers*, and therefore measuring the extent of energy consumption of such devices is not so significant.

In all the *two-tier*, *three-tier* and *four-tier*-based simulations, we have used Epidemic [28], Spray and Wait [29], PRoPHET [30] and MaxProp [31], which are eminent routing protocols for opportunistic networks. We have also used four well-known movement models such as the PDM model [14], *map-based movement model*, *cluster movement model* and *random waypoint model*.

4.2.5.2 Simulation Metrics

We consider four simulation metrics such as *delivery ratio*, *delivery latency*, *energy consumption* and *overhead ratio*. All the metrics except the *delivery ratio* are already defined in Chap. 3 (Sect. 3.3.4.2). Thus, only the *delivery ratio* is defined below.

Delivery ratio: This metric assesses how many numbers of messages reach the destination, out of the total number of messages generated in the network [32]. It is expressed as shown below:

$$\text{Delivery ratio} = \frac{\text{Number of messages delivered}}{\text{Total number of messages generated}}$$

4.2.5.3 Energy Measurement Methodology

The majority of the opportunistic network architectures consist of heterogeneous devices manufactured by different vendors at different hierarchical levels (tiers). In order to make the energy measures uniform, we assume the same type of device is employed in different tiers with minor modifications. Accordingly, we primarily consider smartphones of Samsung Galaxy Grand 2 with Qualcomm Snapdragon processors as the devices for each of the tiers. The other specifications of the devices are Android OS (Jelly Bean) v4.3, 2600 mAh, 3.8 V Li-ion battery. The devices used

Table 4.3 Power consumption of Bluetooth in Samsung Galaxy Grand 2

Bluetooth state	Trepn profiler reading (mW)	Power consumed (mW)	Energy consumption rate (J/s)
Bluetooth off	1284.21	1284.21	1.284
Bluetooth on	1300.73	16.52	0.016
Bluetooth scanning	1540.89	256.68	0.257
Bluetooth pairing	1588.61	304.4	0.305
Bluetooth receiving	1613.46	329.25	0.329
Bluetooth sending	1668.10	383.89	0.384

at *tier-1* are exactly having this specification. However, at the other tiers, the same devices are used with enhanced storage capacity. We measure the communication (e.g. Bluetooth) energy of such devices using the Trepn profiler [33] version 5.0.

During simulation in ONE, various parameters of the *EnergyModel* need to be set. One such parameter is initial stored energy and the other is energy consumption rate for various communication activities (e.g. sending, receiving). The initial stored energy is computed as follows:

As mentioned above, the battery capacity and voltage are 2600 mAh and 3.8 V, respectively. Now, considering 1 Wh = 3600 J and using $E = QV/1000$, we get the initial stored energy as 35568 J.

Further, the energy consumption rate for various activities of Bluetooth communication is measured using the power consumption values of the corresponding activities. The power consumption values are obtained from the respective Trepn profiler readings. In Table 4.3 such power consumption-related values of the specified smartphones are provided for all the Bluetooth communication activities.

4.2.5.4 Results and Discussion

In this section, we analyze the simulation results obtained from different simulation scenarios mentioned in Sect. 4.2.5.1. Four sets of experiments are conducted based on four simulation metrics mentioned in Sect. 4.2.5.2. All the metrics are measured using four types of routing protocols and four types of mobility models, as mentioned in Sect. 4.2.5.1.

In the first set of experiments, the *delivery ratio* is measured in all simulation scenarios mentioned above. Figure 4.3 presents such results. It is evident from the results that the *four-tier* architecture exhibits the best performance in terms of *delivery*

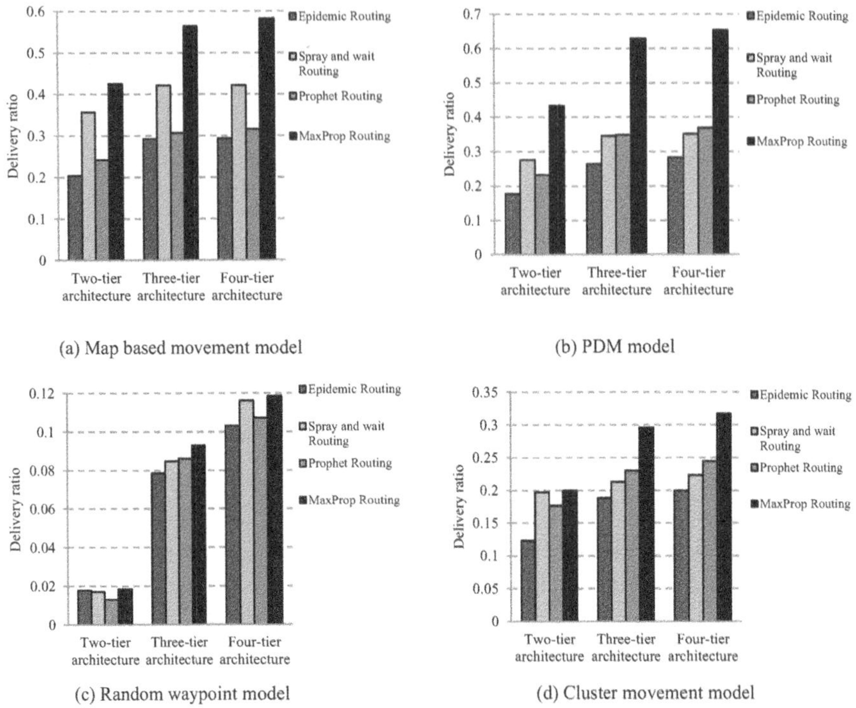

Fig. 4.3 Comparative performance of network architectures in terms of *delivery ratio*

ratio, irrespective of routing protocols and mobility models, whereas *two-tier* architecture exhibits the worst performance. For example, *four-tier* architecture exhibits an average 49.12% and 4.54% improvement in terms of *delivery ratio* compared to *two-tier* and *three-tier* architectures, respectively, under the PDM model. Thus, the improvement made by *four-tier* architecture compared to *three-tier* architecture in terms of the *delivery ratio* is marginal.

In the second set of experiments also, *average delivery latency* is measured in all simulation scenarios mentioned above. In Fig. 4.4, the results of this set of experiments are presented. The experimental results reveal that the *three-tier* and *four-tier* architectures require more time compared to *two-tier* architecture to deliver messages to the destinations irrespective of routing protocols under *map-based movement* and PDM models. On the contrary, *two-tier* architecture requires more time in delivering messages under *random waypoint* and *cluster movement* models. It is further observed that the *four-tier* architecture exhibits marginal improvements compared to *three-tier* architecture in terms of *average delivery latency*. For instance, *four-tier* architecture demonstrates an average 2.21% reduced *average delivery latency* compared to *three-tier* architecture under the *PDM* model.

In the third set of experiments, *average energy consumption* is compared in all the said simulation scenarios. Figure 4.5 depicts such experimental results. The results

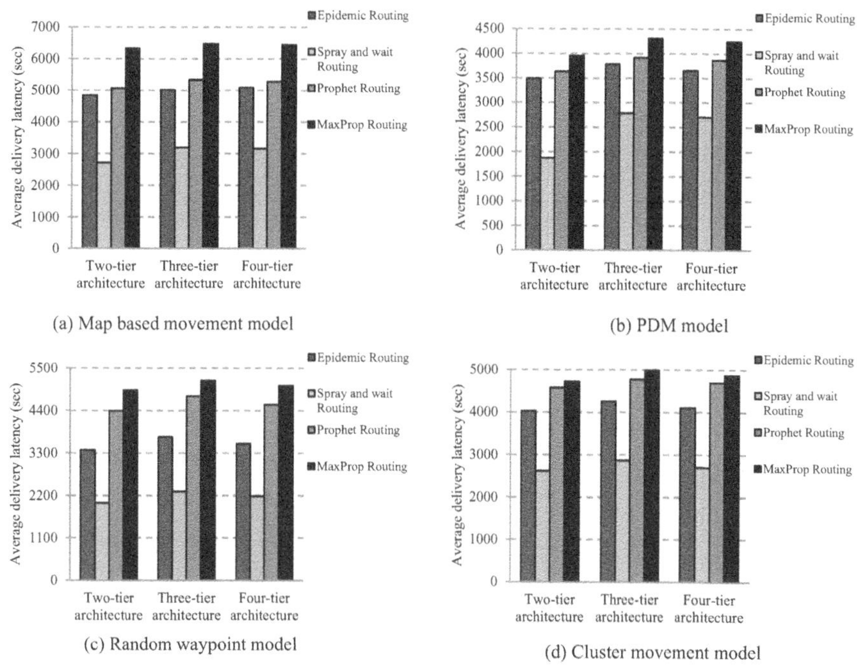

(a) Map based movement model

(b) PDM model

(c) Random waypoint model

(d) Cluster movement model

Fig. 4.4 Comparative performance of network architectures in terms of *average delivery latency*

reveal that *two-tier* consumes additional energy compared to the other two architectures due to the absence of adequate relay nodes (*Throwboxes*). The results also reveal that the *four-tier* architecture consumes marginally lesser energy over the *three-tier* architecture. For instance, *four-tier* architecture demonstrates an average 4.24% reduced *average energy consumption* compared to *three-tier* architecture under the PDM model.

Although the experimental results clearly demonstrate the improvement made by the *four-tier* compared to the *three-tier* architecture in terms of *average energy consumption*, it is to be noted that the energy consumption of *tier-4* devices (e.g. NLOS Wi-Fi towers) is not considered during simulation due to the unavailability of energy consumption data. As mentioned earlier, *tier-4* devices are high-end in terms of energy consumption compared to devices at the other tiers. Precisely, this NLOS Wi-Fi consumes more energy, around four times, compared to Bluetooth [34]. Hence, it can be assumed that *four-tier* architecture consumes more energy compared to *three-tier* architecture.

In the fourth set of experiments, the *overhead ratio* is compared in all the said simulation scenarios. Figure 4.6 depicts such experimental results. Here also we observe that *two-tier* architecture exhibits higher network overhead compared to the other two architectures irrespective of routing protocols and mobility models. It can be further observed from the experimental results that both *three-tier* and *four-tier* architectures exhibit almost identical results in terms of *overhead ratio*. For

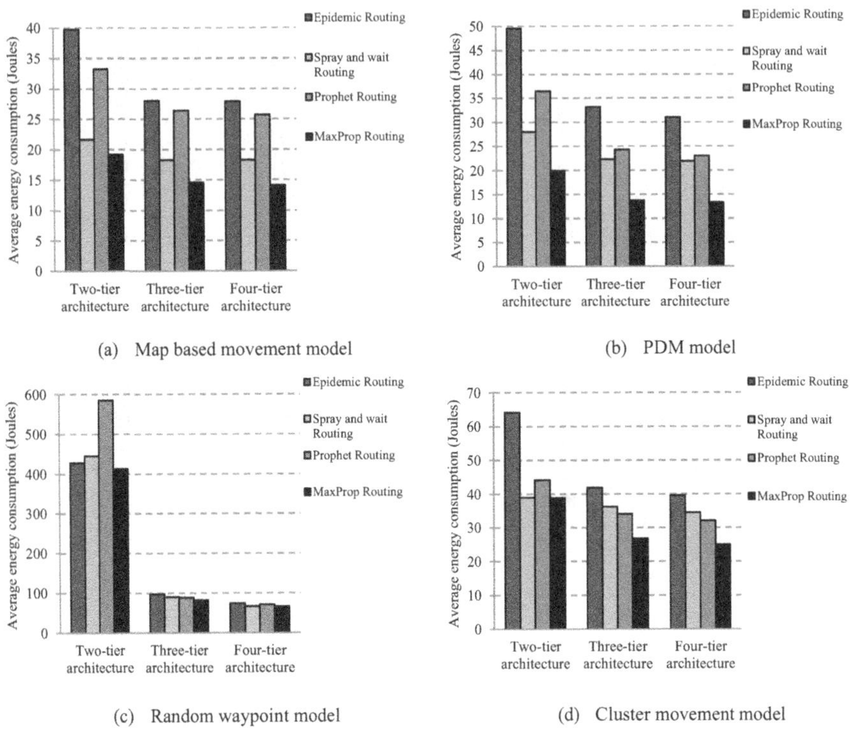

Fig. 4.5 Comparative performance of network architectures in terms of *average energy consumption*

instance, *four-tier* architecture demonstrates an average 0.25% reduced *overhead ratio* compared to *three-tier* architecture under the PDM model.

The aforesaid discussion reveals that the *two-tier* architecture does not perform satisfactorily in terms of design parameters. Hence, *two-tier* architecture proves unsuitable for post-disaster scenarios. The *four-tier* architecture exhibits the best message *delivery ratio*. However, due to the existence of NLOS Wi-Fi towers *four-tier* architecture consumes additional energy compared to the other two architecture categories. Hence, *four-tier* architecture does not satisfy the energy consumption requirement of post-disaster scenarios.

So the implementation of the *three-tier* architecture is recommended in post-disaster scenarios due to its superiority in terms of energy efficiency and almost identical performance compared to *four-tier* architecture in terms of other design parameters. However, if the disaster is too severe then the affected areas get cut off from each other, so using *four-tier* architecture in those areas becomes advantageous.

It is evident from the experimental results presented in Fig. 4.3 that the proposed *three-tier* network architecture with random node deployment can deliver only around 20–63% messages to the destinations successfully. Further, message delivery

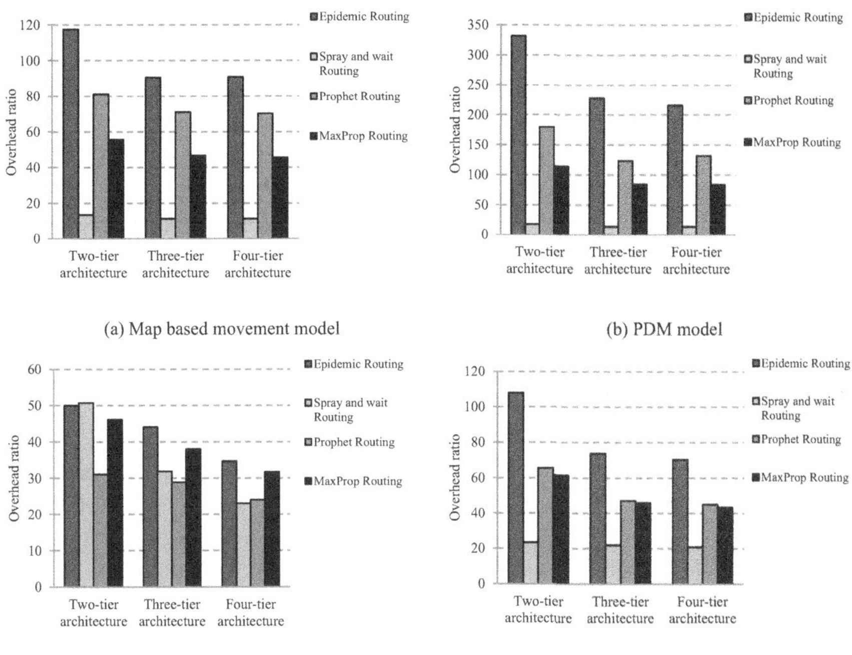

(a) Map based movement model

(b) PDM model

(c)Random waypoint model

(d) Cluster movement model

Fig. 4.6 Comparative performance of network architectures in terms of *overhead ratio*

in opportunistic networks is also dependent on the intra-tier and inter-tier encounters among the nodes. Hence, in order to improve the message delivery further, the appropriate deployment plan of opportunistic network nodes in *tier-1* and *tier-2* proves beneficial. A suitable node deployment for both mobile and static nodes enhances the possibility of encounters among the nodes which improves message delivery and reduces latency, thereby improving network reliability. Hence, in the next section, suitable node deployment plans for both mobile and static nodes are developed in order to enhance the network reliability of post-disaster opportunistic network architecture.

4.3 Node Deployment Plan for Post-disaster Message Transfer

Opportunistic network architecture typically employs a couple of distinct categories of nodes such as mobile nodes and static nodes. The delivery of messages in opportunistic networks is dependent on the encounter among such nodes. However, due to the existence of sporadic connectivity, dissemination of messages on time turns out to be difficult in such networks which results in growing delivery latency. In addition to that, the irregular movement pattern of the nodes causes uncertain encounters which results in unsolicited message drops. Due to such message drops the delivery of messages is hampered. Thus, as a result of the increased latency and reduced message delivery, network reliability is downgraded. In order to enhance network reliability of opportunistic networks, sufficient mobile nodes in *tier-1* and a restricted quantity of stationary nodes (*Throwboxes*) in *tier-2* need to be deployed. The deployment plans for such nodes in post-disaster scenarios are elaborated in the following sections.

4.3.1 Post-disaster Message Dissemination

In a post-disaster scenario, messages (situational information) are usually created at shelters and shared by the *volunteer nodes* with all the other shelters (multicasting) and *camp nodes*. Figure 4.7 illustrates a snapshot of such a message spreading in a scenario containing four shelters, four camps, five TBs and five volunteers.

Here, the nodes' degree of contact opportunity determines the total number of message exchanged among the nodes. On the other hand, the degree of contact opportunity is dependent on the total number of encounters which in turn is dependent on various features such as the number of TBs, nodes' average movement speed and elapsed time after the first deployment of nodes. The volunteers' movement speed and patterns are typically specific to the disaster scenario. Due to this contact opportunity-based environment, the entire set of messages generated at shelters are not exchanged among other shelters and camps within an acceptable delay. Thus, the network reliability is determined by how many messages are successfully arrived at their respective destinations within the acceptable delay.

4.3.2 Mobile Node Deployment

Framing an appropriate plan for deploying mobile nodes in opportunistic networks becomes challenging due to the frequent change in the position of such nodes which makes the opportunistic network's topology highly dynamic. As a result, frequent link failures take place which causes unwanted message drops and poor message delivery. Moreover, in a sparsely populated network where the density of opportunistic network nodes is low, the delay of delivering messages increases. These shortcomings substantially degrade the reliability of a network. In one of our work [35], these limitations are addressed. In the context of opportunistic networks, mobile node deployment refers to fix the quantity of nodes available in a unit area (node density). The total encounters among the nodes become a key parameter to identify suitable node density for reliable data dissemination. Hence, the relationships among the total encounters, node density and other relevant parameters that have an impact

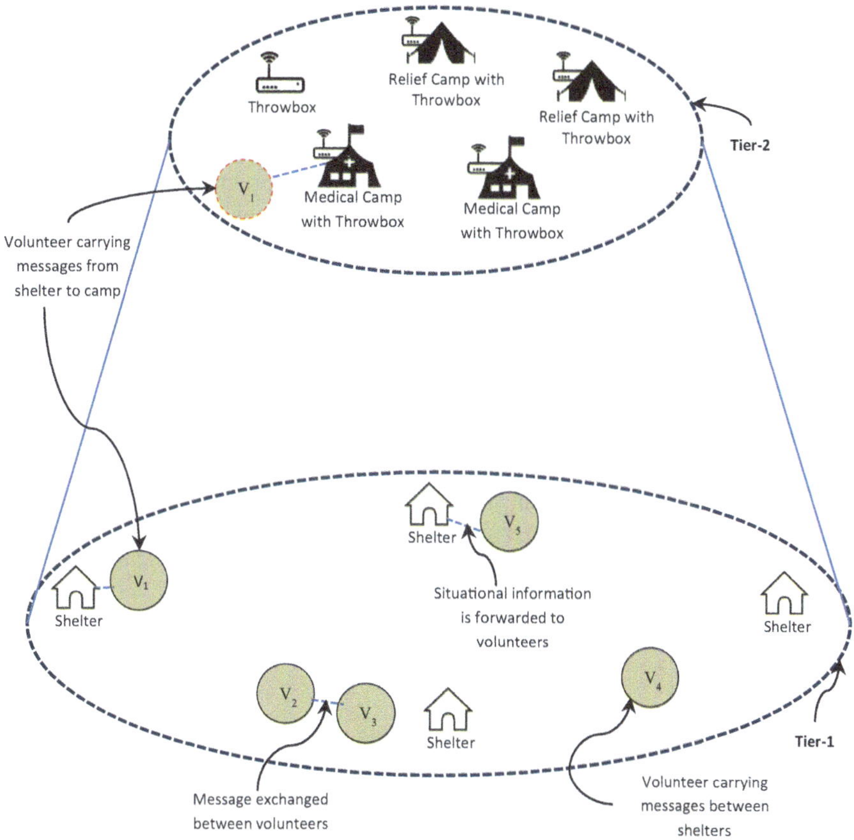

Fig. 4.7 Post-disaster message dissemination involving *tier-1* and *tier-2*

on total encounters such as elapsed time and average movement speed of the mobile nodes are explored. In order to formulate such relationships among the aforesaid parameters, at first, mathematical hypotheses involving such parameters are formulated. Then, the hypotheses are validated through a suitable dataset. The derived relationships after the hypotheses validation assist in mobile node deployment in post-disaster scenarios.

Data dissemination in an opportunistic network is dependent on the contact opportunity among the nodes. The degree of contact opportunity is controlled by the movement pattern of the nodes. In a post-disaster scenario, nodes demonstrate irregular movement patterns that are event-driven in nature, therefore, difficult to model analytically [36]. Hence, the influence of node mobility on the inter-shelter knowledge transfer needs to be quantified. Due to the unavailability of past knowledge regarding the movement patterns of nodes, a metric *total encounters* (E) is identified to assess the degree of contact opportunity based on the comprehensive features of the influence of mobility. The growing *total encounters* among nodes enhance the likelihood of reaching messages to faraway shelters. Hence, improved message delivery is attained. The degree of message spreading over an entire network is assessed by *knowledge sharing ratio* (k) which relies on E. Now, this E is controlled by numerous aspects like node density, the average movement speed of the nodes and total time elapsed since node deployment. This analysis helps to formulate an empirical relationship between the degree of contact opportunity and the degree of knowledge spreading. This relationship assists the response agencies to manage volunteer deployment in affected areas.

4.3.2.1 Analysis of Knowledge Sharing Ratio

This section identifies the parameter(s) which have a substantial impact on the *knowledge sharing ratio*. The amount of knowledge sharing can be forecasted by observing these parameter values.

As per the aforesaid discussion an escalation in *total encounters* causes the likelihood of delivering knowledge to remote shelters. As a result, the *knowledge sharing ratio* (k) improves. The *total encounters* (E) can be expressed as

$$E = \sum_{t=1}^{n} c_t \tag{4.1}$$

where c_t is the number of meetings among nodes with t duration, t is the time length of each meeting, and n is the quantity of all probable distinct meeting durations.

Therefore, from the above discussion, the following is derived:

$$k = f_1(E) \tag{4.2}$$

Total encounters is expressed in Eq. 4.1 using the number of meetings among nodes (c_t). The movement pattern of the nodes primarily influences c_t. The movement pattern can be expressed in terms of speed and direction of movement. Due to the event-driven nature, it is nearly impossible to forecast the movement direction of nodes in the post-disaster situation. Hence, for simplicity, we assume that c_t relies on the average movement speed of the nodes symbolized by s. In addition to that, c_t also relies on node density (nd) and elapsed time (T). The nd includes both static and mobile nodes. However, in the context of mobile node deployment, nd implies the quantity of *volunteer nodes* positioned in a unit area only, while the number of static nodes is kept as constant. If nd in an area is increased by keeping T and s as constants, c_t increases. Similarly, c_t increases with T by keeping nd and s constants. Hence, c_t can be expressed as a combination of nd, T and s, and consequently, E can also be denoted as

$$E = f_2(nd, T, s) \tag{4.3}$$

Substituting E in Eq. 4.2, the following is derived

$$k = f_3(nd, T, s) \tag{4.4}$$

The relationship in Eq. 4.4 provides a recommendation regarding mobile node deployment.

4.3.2.2 Modeling the Deployment Plan

Here we provide a mathematically devised hypothesis involving *total encounters* and *knowledge sharing ratio*. To devise such a hypothesis, we simulate instances based on the post-disaster dataset collected after the Nepal earthquake on 25 April 2015 and analyze the outcome of the simulation. Next, we validate the hypothesis using another simulation outcome. This outcome is acquired by simulating instances using Namkhana field trial based volunteers' mobility traces. This field trial has been conducted on 27 and 28 February 2016 at Namkhana which is a disaster-prone locality in West Bengal, India. The simulation environment of the Nepal earthquake is configured using the information received from several sources [37, 38]. The details are already provided in Chap. 3 (Sect. 3.3.4.1). However, the details of Namkhana field trial-based simulation are elaborated below.

We have identified four shelters in 1 km^2 area in the region of Ramakrishna Ashram located at Chandanpiri village at Namkhana Block for mock field trials. In this field trial, we have deployed 37 volunteers carrying smartphones/laptops. But we could collect the traces of 34 volunteers whereas 3 volunteers' traces could not be collected as their area of movement was beyond the network coverage zone. Thus, in ONE simulator these 34 volunteers' traces have been fed.

Hypothesis Formulation

The experimental results acquired from the simulation of the Nepal earthquake dataset are plotted in Fig. 4.8. The impact of *total encounters* on *knowledge sharing ratio* is shown in Fig. 4.8a. We observe that after a certain number of total encounters the plot starts to rise sharply and at last reaches the steady-state. The reason for this nature of the plot may be explained like this. In the simulation, to start with, the nodes near the shelters take the situational messages from the shelters and exchange the same with the adjacent nodes. So, at this stage, the situational messages are not delivered to the destinations, and as a result, *knowledge sharing ratio* does not rise significantly. However, over time, encounters increase and the nodes which collect situational messages slowly would deliver the same to the destinations. This increases the *knowledge sharing ratio*. The nature of the plot *knowledge sharing ratio* versus *total encounters* conforms to the nature of the logistic "S" shape curve. Thus, the hypothesis for this scenario is:

$$k = \frac{L}{1 + e^{-\lambda(E-M)}}$$

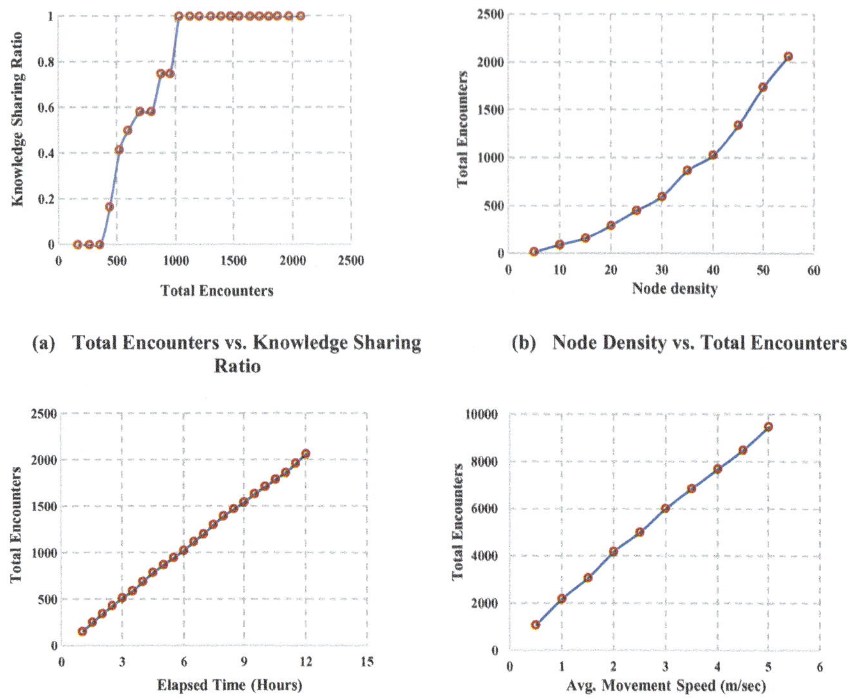

(a) Total Encounters vs. Knowledge Sharing Ratio

(b) Node Density vs. Total Encounters

(c) Elapsed Time vs. Total Encounters

(d) Average movement speed vs. Total Encounters

Fig. 4.8 Inter-relationship between different parameters in the Nepal disaster

Here, k and E represent *knowledge sharing ratio* and *total encounters*, respectively. Also, L and λ represent the peak and slope of the curve, whereas M represents E value at the middle point of the curve. In our application scenario, L would be 1. Thus, the above expression is modified as follows:

$$k = \frac{1}{1 + e^{-\lambda(E-M)}} \tag{4.5}$$

Figure 4.8b demonstrates how *total encounters* is influenced by node density during the post-disaster scenario after the Nepal earthquake. It is observed from the figure that as node density increases *total encounters* grows exponentially. So the hypothesis is formulated as follows:

$$E = a_1 nd^{b_1} \tag{4.6}$$

where nd is the node density, and a_1 and b_1 are constants. Similarly, Fig. 4.8c illustrates the influence of elapsed time on *total encounters*. Our hypothesis for this scenario is

$$E = a_2 + b_2 T \tag{4.7}$$

where T is the elapsed time, and a_2 and b_2 are constants.

Figure 4.8d exhibits the influence of average movement speed on *total encounters*. So, the hypothesis for this scenario is

$$E = a_3 + b_3 s \tag{4.8}$$

where s is the average movement speed of nodes, a_3 and b_3 are constants.

Hypothesis Validation and Parameter Estimation

To validate the proposed hypotheses, the results acquired from simulation based on Namkhana field trial volunteers' traces are plotted in Fig. 4.9. We observe from Fig. 4.9a that the nature of the plot is similar to Fig. 4.8a. Thus, it validates the hypothesis as in Eq. 4.5.

By fitting an appropriate curve on the dataset obtained from Fig. 4.9a, the following is derived

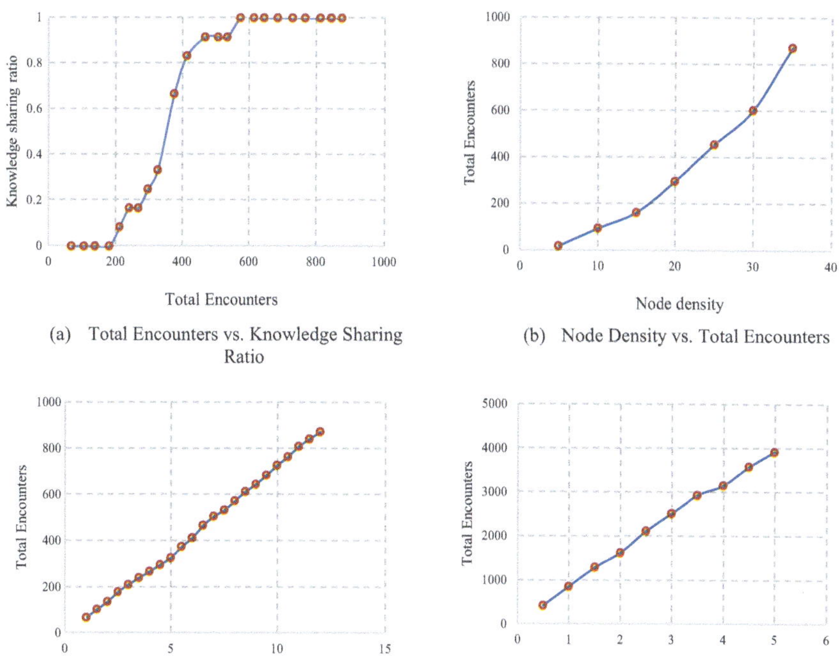

(a) Total Encounters vs. Knowledge Sharing Ratio

(b) Node Density vs. Total Encounters

(c) Elapsed Time vs. Total Encounters

(d) Average movement speed vs. Total Encounters

Fig. 4.9 Inter-relationship between different parameters in the Namkhana field trial

Table 4.4 Estimated values of constants along with accuracy

Equation	Constant	Value	Percentage error (%)
$E = a_1 nd^{b_1}$	a_1	1.12	5.71
	b_1	1.86	
$E = a_2 + b_2 T$	a_2	−18.41	3.42
	b_2	74.33	
$E = a_3 + b_3 s$	a_3	124.13	3.72
	b_3	776.31	

$$k = \frac{1}{1 + e^{-0.02(E-348)}} \tag{4.9}$$

Further, to evaluate the accuracy of the k obtained as above, we compute RMSE (root mean square error) and PE (percentage error) from the results plotted in Fig. 4.9a. The RMSE and PE are computed as 0.0286 and 6.05%, respectively.

Similar to the above validation, the results acquired from simulation based on Namkhana field trial volunteers' traces are plotted in Fig. 4.9. It shows how *total encounters* is impacted by node density. We observe from Fig. 4.9b that the nature of the plot is similar to Fig. 4.8b. Thus, it validates the hypothesis as in Eq. 4.6.

Similar to the above validation, we validate the hypotheses in Eqs. (4.7) and (4.8) with the help of the results plotted in Fig. 4.9c, d, respectively.

Next, to compute the constant values (a_1, b_1), (a_2, b_2) and (a_3, b_3), we employ the least square method on the results of Fig. 4.9b–d, respectively. Table 4.4 presents such constant values and their respective errors in percentage. We notice that the range of errors is 3.4–5.7%, which is tolerable in our post-disaster application domain.

Now, combining Eqs. (4.6), (4.7) and (4.8) the following can be written:

$$E = ua_1 nd^{b_1}(a_3 + b_3 s)(a_2 + b_2 T) \tag{4.10}$$

Here, u represents a constant which is calculated using the results acquired from the Namkhana-based field trial. The estimated u values belong to the range $2.226 \times 10^{-6} \leq u \leq 2.338 \times 10^{-6}$. We consider the highest value of u in computing the parameters wherever applicable. Now, by substituting the values of E and u in Eq. (4.9) the following is derived:

$$k = \frac{1}{1 + e^{-0.02\{2.338\times 10^{-6} \times a_1 nd^{b_1}(a_3+b_3 s)(a_2+b_2 T)-348\}}} \tag{4.11}$$

Equation 4.11 represents a relationship between k, nd, T and s.

We observe from Eq. 4.11 that we can estimate the attainable value of *knowledge sharing ratio* using the values of nd, T and s. Similarly, the required node density can be predicted if the values of k, T and s are known. In this manner, if any three variables are known out of k, nd, T and s, the rest can be calculated. This derivation helps in volunteer deployment in post-disaster scenarios.

4.3.2.3 Performance Evaluation

This section evaluates the performance of the proposed mobile node deployment plan through simulation in ONE simulator. Due to the unavailability of suitable competing schemes for mobile node deployment, the performance of the proposed mobile node deployment plan cannot be compared.

Simulation Setup

The simulation setup is identical to the setup mentioned in Sect. 3.2.2.2. Two well-known opportunistic network routing protocols such as PRoPHET and Epidemic routing protocols are used for simulation. During the simulation, the number of TBs (five) is kept constant. The node density (the number of *volunteer nodes* in the unit area) is increased gradually.

Simulation Metrics

During simulation four metrics, such as *overhead ratio, average delivery latency, average energy consumption* and *knowledge sharing ratio*, are considered for performance evaluation. Out of these the first three are already defined in Sect. 3.1.5.2. The additional metric *knowledge sharing ratio* is defined below.

Knowledge sharing ratio (k): It specifies how much situational message is spread in the whole network within a pre-determined time period [35]. We define this metric as follows:

$$k = \frac{\text{Average number of situational messages received}}{\text{Total number of situational messages generated}}$$

Results and Discussion

We conduct two sets of experiments through which our mobile node deployment plan would be estimated from the perspective of network reliability. We estimate how *knowledge sharing ratio* varies with node density through the first set of experiments and Fig. 4.10a presents the results.

The figure indicates that for both the routing protocols, the *knowledge sharing ratio* grows steadily with increasing node density and progressively reaches a steady state. It is observed from the experimental results that, employing node density 25 for the simulation period of 12 h, the values of *knowledge sharing ratio* are 0.866 and 0.85 in PRoPHET and Epidemic routing protocols, respectively. Increasing the node density further does not improve the *knowledge sharing ratio* substantially.

Now, using Eq. 4.11 the node density required to achieve a desired *knowledge sharing ratio* within a stipulated interval can be predicted theoretically when the average movement speed of the nodes is known. Assuming the average movement speed of the volunteers is 0.5 m/s (average walking speed), the node density required to achieve *knowledge sharing ratio* 0.90 within 12 h of node deployment is estimated as 24.7 (approximately 25). It is observed from the experimental results that using the

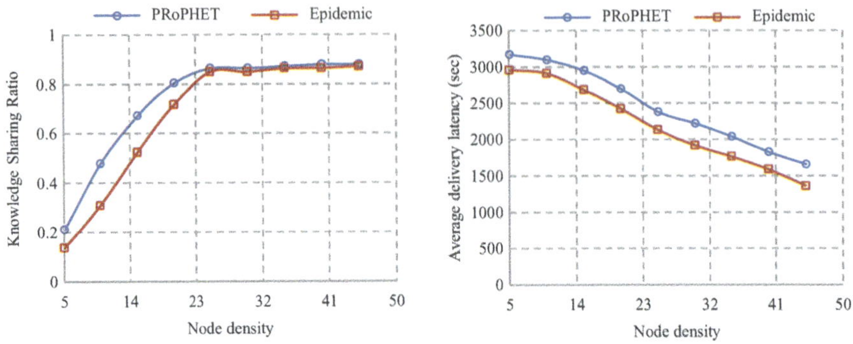

(a) Knowledge Sharing Ratio (b) Average delivery latency

Fig. 4.10 Comparative performance of the proposed mobile node deployment plan

aforesaid values of different parameters, the achievable values of *knowledge sharing ratio* are 0.866 and 0.85 in PRoPHET and Epidemic routing protocols, respectively. Hence, the theoretical estimation of the *knowledge sharing ratio* using Eq. 4.11 contains 4.65 and 5.88% errors for the two routing protocols.

We estimate how the average delivery latency varies with node density through the second set of experiments and Fig. 4.10b presents the results. The figure clearly reveals that the average delivery latency decreases almost linearly with increasing node density for both the routing protocols. Hence the network reliability is enhanced.

Summarily, it is observed from the first and second sets of experiments that the proposed mobile node deployment plan enhances network reliability and estimates node density with acceptable accuracy (within 3–6% error). Thus, the proposed mobile node deployment plan can be used by disaster response agencies in volunteer (mobile node) deployment during post-disaster scenarios. However, in post-disaster scenarios volunteers are scarce entities. Hence, increasing the number of *volunteer nodes* in *tier-1* of the opportunistic networks-based post-disaster network architecture to improve network reliability proves infeasible in such scenarios. Such a deficit of mobile nodes can be compensated by deploying limited quantities of static nodes in *tier-2*. Thus, in the next section, a static node deployment plan is elaborated.

4.3.3 Static Node Deployment

The proposed deployment plan for mobile nodes is explained in Sect. 4.3.2. In this section, a suitable deployment plan for static nodes (*Throwboxes*) is described. The appropriate placements of TBs produce added meeting opportunities among the nodes which improve network reliability. In order to place TBs efficiently, previous information concerning network parameters like the scale of the network, topology and movement pattern of nodes, etc. is crucial. For instance, placement of TBs in places visited by volunteers occasionally does not enhance network reliability. Likewise, placing a huge quantity of TBs in random locations only escalates the maintenance cost of the network. Thus, an appropriate *Throwbox* placement strategy in post-disaster scenarios necessitates techniques for identifying the required quantity of TBs and identifying suitable locations for deploying them. Although the availability of prior information regarding network parameters in post-disaster scenarios becomes challenging, hence, an encounter-based *Throwbox* deployment strategy (EnTER) [39] is described in the following section. In order to determine the required number of TBs, the EnTER strategy employs a similar technique of calculating the number of mobile nodes in a unit area presented in Sect. 4.3.2.2.

4.3.3.1 Proposed EnTER Strategy

This section illustrates the EnTER strategy which involves two sub-tasks such as identifying the necessary quantity of TBs and identifying the appropriate positions for deploying such TBs for improving network reliability.

Determining Required Number of TBs

It has been observed from Sect. 4.3.2.1 that node density (*nd*) in opportunistic networks includes both static and mobile nodes. During mobile node deployment, *nd* is solely reliant on the number of mobile nodes in a unit area while keeping the quantity of static nodes as constant. Thus, in the context of static node deployment, *nd* implies the number of TBs (*ntb*) deployed in a unit area, while the quantity of mobile nodes is kept constant. Hence, the derived equation obtained from Eq. (4.11) for identifying the required number of TBs is

$$k = \frac{1}{1 + e^{-0.02\{2.338\times 10^{-6}\times a_1 ntb^{b_1}(a_3+b_3 s)(a_2+b_2 T)-348\}}} \tag{4.12}$$

The estimated values of the other constants along with the corresponding percentage errors are identical to the values in Table 4.4. From Eq. 4.12, the attainable *knowledge sharing ratio* can be forecasted if the values of *ntb*, T and s are known. Similarly, *ntb* can be predicted if the values of k, T and s are known in advance. Likewise, if the values of any three variables are known out of the four (k, *ntb*, T and s) the rest can be obtained. This derivation facilitates the estimation of the necessary number of TBs in post-disaster scenarios.

Identifying Suitable Locations for TB Deployment

The *strategic placement* of TBs is employed in the proposed EnTER strategy as the *random placement* strategy does not yield the desired network reliability. In the disaster aftermath, the availability of information regarding network parameters like mobility characteristics and encounter patterns of the nodes, etc. in advance is ruled out. Thus, a hybrid strategy for TB placement is introduced. According to the strategy, the TBs are positioned initially at camps. Further, due to the spatial regularity characteristic of human mobility, the volunteers tend to return to their respective camps periodically. Hence, such deployment increases the possibility of situational information dissemination among the volunteers. Later, upon the availability of information regarding network parameters, the TBs are positioned in the most regularly visited places by volunteers. Such locations in the affected areas are called the *most popular locations*. The location i with the highest degree centrality is designated as the *most popular location*. The degree centrality of location i, ($D(i)$), usually specifies the quantity of volunteers considering i as one of their top locations. Greater $D(i)$ signifies i is more popular among volunteers.

Here, $D(i)$ is calculated based on the concept of *Heatmap* as mentioned in [40]. A *Heatmap* is a representation of the relative distribution of volunteers on the entire two-dimensional disaster-affected area. The entire disaster-affected area is divided into a small block (zone) of equal size. Every volunteer maintains a *Heatmap* in the form of a matrix in which each cell corresponds to a zone on the map of the disaster-affected area. The value in each cell of the matrix is assigned depending on the number of volunteers visiting that cell. The volunteers exchange this matrix during their encounter with other volunteers and update their own *Heatmap*. The value of each cell of the matrix indicates the degree centrality of a location. Hence, from the *Heatmap* of volunteers, the *most popular locations* in the disaster-affected areas can be identified. Now, in order to deploy TBs in the *most popular locations*, three scenarios might occur as follows:

- Number of TBs is equal to the number of *most popular locations*
- Number of TBs is greater than the number of *most popular locations*
- Number of TBs is less than the number of *most popular locations*

In the first two scenarios, we simply deploy TBs in the *most popular locations*. However, in the third scenario, TBs are deployed based on the higher values of $D(i)$. In the case of multiple locations having the same $D(i)$ value, priority is given to the location having the smallest Euclidean distance from a shelter.

4.3.3.2 Performance Evaluation

The performance of the proposed static node deployment plan (EnTER strategy) is evaluated through simulation on ONE simulator in this section.

Simulation Setup

The simulation arrangement is identical to Sect. 4.3.2.2. The other parameters such as the mobility models and network architecture are also identical to Sect. 4.3.2.2. Here, the Epidemic routing protocol is used for the simulation. However, due to the scarcity of volunteers in post-disaster scenarios, the number of *volunteer nodes* is set to 60 (constant) throughout the simulation whereas the quantity of TBs is increased gradually.

Simulation Metrics

Four metrics such as *knowledge sharing ratio*, *overhead ratio*, *average delivery latency* and *average energy consumption* are considered as simulation metrics. The *knowledge sharing ratio* is defined in Sect. 4.3.2.3 whereas the other three metrics are defined in Sect. 4.2.5.2.

Results and Discussion

To assess the EnTER, our *Throwbox* deployment approach, we conduct four sets of experiments for a 12-h simulation period. Performance is assessed in terms of network reliability through the first two sets of experiments. On the other hand, the feasibility of the EnTER is assessed through the other two sets of experiments. As the nodes are energy-starved and recharging may not be possible whenever required, the feasibility assessment is important in terms of energy consumption and overheads. In all sets of experiments, we compare the EnTER with one state-of-the-art TB placement approach RPLM [25].

Figure 4.11a presents the results of the first set of experiments where we estimate how the *knowledge sharing ratio* changes with varying number of TBs. The results

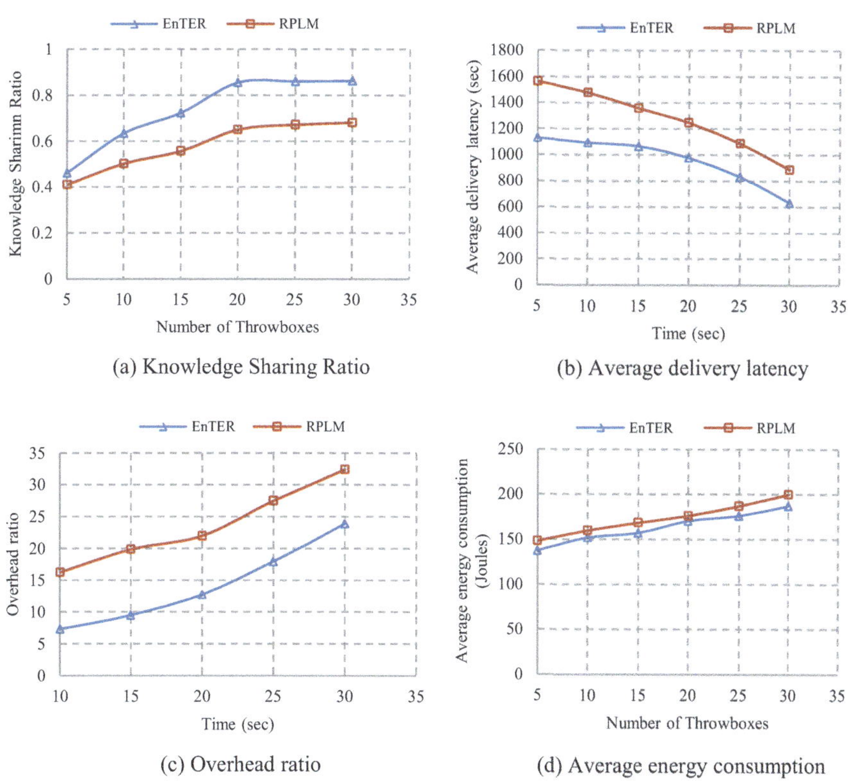

Fig. 4.11 Comparative performance of EnTER strategy

indicate that for both the EnTER and RPLM, as the number of TBs increases *knowledge sharing ratio* increases gradually and finally reaches the steady-state. It also indicates that the EnTER shows around 25.78% higher *knowledge sharing ratio* over the RPLM. For example, for 25 TBs the values of *knowledge sharing ratio* are 0.862 and 0.673 in EnTER and RPLM, respectively. Here the improvement is 28.08%.

Figure 4.11b presents the results of the second set of experiments where we estimate how the average delivery latency changes with varying number of TBs. The results indicate that as the number of TBs increases average delivery latency falls more or less linearly for both the competing strategies. The EnTER shows around 33.12% less delivery latency over the RPLM. To be more specific, for 25 TBs deployed, the average delivery latencies are 832.71.23 s and 1086.81 s in EnTER and RPLM, respectively. Summarily, from the first two sets of experiments, it is observed that the EnTER exhibits higher network reliability over the RPLM from the perspective of both *knowledge sharing ratio* and average delivery latency.

We assess the *overhead ratio* with varying number of TBs for both the EnTER and RPLM through the third set of experiments and plot the results in Fig. 4.11c. It is observed from the plots that for both the schemes, the *overhead ratio* increases with the number of TBs. However, in EnTER always the *overhead ratio* is less than RPLM. For example, for 25 TBs deployed, the *overhead ratios* are 18.03 and 27.53 in EnTER and RPLM, respectively. Further, EnTER exhibits an average 44.20% reduced *overhead ratio* compared to RPLM which indicates the transmission cost of messages in EnTER is comparatively less than RPLM.

We estimate the average energy consumption with a varying number of TBs for both the EnTER and RPLM through the fourth set of experiments and plot the results in Fig. 4.11d. We observe from the plots that for both the schemes, the *average energy consumption* increases with the number of TBs. The results reveal the marginal improvement of the EnTER over RPLM in terms of *average energy consumption*. For instance, for 25 TBs deployed, in EnTER average energy consumed is approximately 176.09 J, whereas in RPLM this value is approximately 186.77 J. Hence, the proposed EnTER strategy is marginally better than RPLM in terms of average energy consumption.

4.4 Conclusion

In this chapter, the design of reliable post-disaster opportunistic network architecture is discussed. The proposed opportunistic network architecture includes a *three-tier* network architecture and node deployment plans involving mobile and static nodes in order to provide reliable communication in the disaster aftermath. However, performance evaluation of the proposed *three-tier* network architecture indicates that the message up to 63% can be delivered by implementing such architecture. The message delivery in opportunistic networks is dependent on the intra-tier and inter-tier encounters among the nodes. A suitable node deployment in both *tier-1* and *tier-2* enhances the possibility of encounters among the nodes which improves message delivery and reduces latency, thereby improving network reliability. Thus, in order to improve the performance further, appropriate node deployment plans for both mobile and static nodes are also proposed, which build the opportunistic network architecture.

It has been observed that the introduction of an appropriate node deployment plan for both static and mobile nodes improves the message delivery up to 86% (*knowledge sharing ratio* is 0.86) with average delivery latency of 631.23 s. Routing has a great role to play in message delivery. This motivates us to discuss appropriate opportunistic network routing protocols toward improving the said performance in providing post-disaster management services in the next chapter.

References

1. Fall, K., Iannaccone, G., Kannan, J., Silveira, F., Taft, N.: A Disruption-Tolerant Architecture for Secure and Efficient Disaster Response Communications. https://www.semanticscholar.org/paper/A-Disruption-Tolerant-Architecture-for-Secure-and-Fall-Iannaccone/a4340ff8e0e5a9a03b06dd77f5984ea25778001a (2010)
2. Ntareme, H., Zennaro, M., Pehrson, B.: Delay tolerant network on smartphones. In: Proceedings of the 3rd Extreme Conference on Communication The Amazon Expedition—ExtremeCom '11, pp. 1–6. ACM Press, New York, NY, USA (2011)
3. Paul, P.S., Nandi, S., Dey, S.K., De, K., Pramanik, P., Saha, S.: Challenges in designing testbed for evaluating delay-tolerant hybrid networks. In: 2015 IEEE International Conference on Pervasive Computing and Communication Workshops (PerCom Workshops). pp. 280–283. IEEE (2015)
4. Dilmaghani, R., Rao, R.: A wireless mesh infrastructure deployment with application for emergency scenarios. In: Proceedings of the 5th International ISCRAM Conference, pp. 1–11 (2008). https://doi.org/10.1099/vir.0.068536-0
5. George, S., Zhou, W., Chenji, H., Won, M., Lee, Y., Pazarloglou, A., Stoleru, R., Barooah, P.: DistressNet: a wireless ad hoc and sensor network architecture for situation management in disaster response. IEEE Commun. Mag. **48**, 128–136 (2010). https://doi.org/10.1109/MCOM.2010.5434384
6. Chenji, H., Chenji, H., Hassanzadeh, A., Won, M., Li, Y., Zhang, W., Yang, X., Stoleru, R., Zhou, G.: A Wireless Sensor, AdHoc and Delay Tolerant Network System for Disaster Response
7. Saha, S., Shah, V.K., Verma, R., Mandal, R., Nandi, S.: Is it worth taking a planned approach to design ad hoc infrastructure for post disaster communication? In: Proceedings of the Seventh ACM International Workshop on Challenged Networks—CHANTS '12, p. 87 (2012). https://doi.org/10.1145/2348616.2348637
8. Saha, S., Nandi, S., Paul, P.S., Shah, V.K., Roy, A., Das, S.K.: Designing delay constrained hybrid ad hoc network infrastructure for post-disaster communication. Ad Hoc Netw. **25**, 406–429 (2015). https://doi.org/10.1016/j.adhoc.2014.08.009
9. Gupta, A.K., Bhattacharya, I., Banerjee, P.S., Mandal, J.K., Mukherjee, A.: DirMove: direction of movement based routing in DTN architecture for post-disaster scenario. Wirel. Netw. **22**, 723–740 (2016). https://doi.org/10.1007/s11276-015-0994-0
10. Bhattacharjee, S., Roy, S., Bandyopadhyay, S.: Exploring an energy-efficient DTN framework supporting disaster management services in post disaster relief operation. Wirel. Netw. **21** (2015). https://doi.org/10.1007/s11276-014-0836-5
11. Banerjee, N., Corner, M.D., Levine, B.N.: An Energy-efficient architecture for DTN throwboxes. In: Proceedings of the IEEE INFOCOM, pp. 776–784 (2007). https://doi.org/10.1109/INFCOM.2007.96

12. Uchida, N., Kawamura, N., Ishida, T., Shibata, Y.: Proposal of seeking wireless station by flight drones based on delay tolerant networks. In: 2014 Ninth International Conference on Broadband and Wireless Computing, Communication and Applications, pp. 401–405. IEEE (2014)
13. Pentland, A., Fletcher, R., Hasson, A.: DakNet: rethinking connectivity in developing nations. Computer (Long Beach, California) **37**, 78–83 (2004). https://doi.org/10.1109/MC.2004.1260729
14. Uddin, Y.S., Nicol, D.M., Abdelzaher, T.F., Kravets, R.H.: A post-disaster mobility model for delay tolerant networking. In: Proceedings of the Winter Simulation Conference, pp. 2785–2796 (2009). https://doi.org/10.1109/WSC.2009.5429249
15. De Pellegrini, F., Miorandi, D., Carreras, I., Chlamtac, I.: A graph-based model for disconnected ad hoc networks. In: IEEE INFOCOM 2007—26th IEEE International Conference on Computer Communications, pp. 373–381. IEEE (2007)
16. Ogura, M., Preciado, V.M.: Stability of spreading processes over time-varying large-scale networks. IEEE Trans. Netw. Sci. Eng. **3**, 44–57 (2016). https://doi.org/10.1109/TNSE.2016.2516346
17. de Pebeyre, A.F., Tarissan, F., Sopena, J.: On the relevance of the edge-Markovian evolving graph model for real mobile networks. In: 2013 IFIP Wireless Days (WD), pp. 1–6. IEEE (2013)
18. Ding, Y., Li, N., Song, B., Yang, Y.: The mobile node deployment algorithm for underwater wireless sensor networks. In: 2017 Chinese Automation Congress (CAC), pp. 456–460. IEEE (2017)
19. Roselin, J., Latha, P.: Prioritized mobile node deployment to enhance the lifetime of wireless sensor network. Wirel. Pers. Commun. **96**, 3693–3711 (2017). https://doi.org/10.1007/s11277-017-4062-1
20. Qirtas, M.M., Faheem, Y., Rehmani, M.H.: Throwboxes in delay tolerant networks: a survey of placement strategies, buffering capacity, and mobility models. J. Netw. Comput. Appl. **91**, 89–103 (2017). https://doi.org/10.1016/J.JNCA.2017.04.004
21. Xiao, M., Wu, J., Huang, L.: Home-based zero-knowledge multi-copy routing in mobile social networks. IEEE Trans. Parallel Distrib. Syst. **26**, 1238–1250 (2015). https://doi.org/10.1109/TPDS.2014.2319211
22. You, L., Li, J., Wei, C., Hu, L.: MPAR: a movement pattern-aware optimal routing for social delay tolerant networks. Ad Hoc Netw. **24**, 228–249 (2015). https://doi.org/10.1016/J.ADHOC.2014.09.004
23. Li, F., Yin, Z., Tang, S., Zhang, C., Cheng, Y., Wang, Y.: K-throwbox placement problem in throwbox-assisted delay tolerant networks. In: 2014 IEEE Global Communications Conference, pp. 253–258. IEEE (2014)
24. Zhu, Y., Zhang, C., Mao, X., Wang, Y.: Social based throwbox placement schemes for large-scale mobile social delay tolerant networks. Comput. Commun. **65**, 10–26 (2015). https://doi.org/10.1016/J.COMCOM.2015.02.008
25. Li, P., Kawasaki, T., Miyazaki, T., Guo, S.: Relay placement for latency minimization in delay tolerant networks. In: 2015 IEEE International Conference on Communications (ICC), pp. 3819–3824. IEEE (2015)
26. Keränen, A., Ott, J., Kärkkäinen, T.: The ONE simulator for DTN protocol evaluation. In: Proceedings of the Second International ICST Conference on Simulation Tools and Techniques, p. 55. ICST (2009)
27. Uttarakhand Flooding. http://google.org/crisismap/2013-uttrakhand-floods?gl=in
28. Vogels, W., van Renesse, R., Birman, K.: The power of epidemics. ACM SIGCOMM Comput. Commun. Rev. **33**, 131–135 (2003). https://doi.org/10.1145/774763.774784
29. Spyropoulos, T., Psounis, K., Raghavendra, C.S.: Spray and wait: an efficient routing scheme for intermittently connected mobile networks. In: Proceedings of the 2005 ACM SIGCOMM Workshop on Delay-Tolerant Networking—WDTN '05, pp. 252–259 (2005). https://doi.org/10.1145/1080139.1080143

30. Lindgren, A., Doria, A., Schelén, O.: Probabilistic routing in intermittently connected networks. ACM SIGMOBILE Mob. Comput. Commun. Rev. **7**, 19 (2003). https://doi.org/10.1145/961268.961272
31. Burgess, J., Gallagher, B., Jensen, D., Levine, B.N.: MaxProp: routing for vehicle-based disruption-tolerant networks. In: Proceedings of the IEEE INFOCOM 2006. 25TH IEEE International Conference on Computer Communications, pp. 1–11. IEEE (2006)
32. Misra, S., Saha, B.K., Pal, S.: Opportunistic mobile networks: advances and applications
33. Ben-Zur, L.: Trepn Power Profiler—Qualcomm Developer Network. https://developer.qualcomm.com/software/trepn-power-profiler
34. Perrucci, G.P., Fitzek, F.H.P., Widmer, J.: Survey on energy consumption entities on the smartphone platform. In: 2011 IEEE 73rd Vehicular Technology Conference (VTC Spring), pp. 1–6. IEEE (2011)
35. Bhattacharjee, S., Roy, S., Ghosh, S., DasBit, S.: Exploring the impact of connectivity on dissemination of post disaster situational data over DTN. In: Proceedings of the 18th International Conference on Distributed Computing and Networking—ICDCN '17, pp. 1–4. ACM Press, New York, NY, USA (2017)
36. Picu, A., Spyropoulos, T.: DTN-Meteo: forecasting the performance of DTN protocols under heterogeneous mobility. IEEE/ACM Trans. Netw. **23**, 587–602 (2015). https://doi.org/10.1109/TNET.2014.2301376
37. Nepal Quake Unofficial Crisis Map—Google My Maps. https://www.google.com/maps/d/viewer?mid=1_PvxC6bAC9zpk86jRHvAuLCPJ3k&ll=27.692258449090865%2C85.33920080859377&z=10
38. Learning from the April 25, 2015, Nepal earthquake: Mapping the deformation and site response, Bangalore, India
39. Bhattacharjee, S., Bit, S. Das: EnTER: an encounter based throwbox deployment strategy for enhancing network reliability in post-disaster scenarios over DTN. In: Proceedings of the 20th International Conference on Distributed Computing and Networking—ICDCN '19, pp. 413–416. ACM Press, New York, NY, USA (2019)
40. Mukherjee, A., Basu, S., Roy, S., Bandyopadhyay, S.: Developing a coherent global view for post disaster situation awareness using opportunistic network. In: 2015 7th International Conference on Communication Systems and Networks (COMSNETS), pp. 1–8. IEEE (2015)

Chapter 5
Energy-Efficient Post-disaster Routing Protocols

In this chapter, different energy-efficient unicast and multicast opportunistic network routing protocols to be used for different post-disaster management services are discussed. Some services such as situational awareness primarily rely on delivering messages to single or multiple destinations, whereas some services typically require a single message to be forwarded to a group of pre-defined nodes. This implies that the services require unicast and/or multicast communication. So, three routing protocols, viz., Priority-enhanced PRoPHET (Pen-PRoPHET), Group encounter-based PRoPHET (Ge-PRoPHET) and Trace Route are proposed supporting different post-disaster management services.

The primary aim of proposing opportunistic network routing protocols for post-disaster communication is to increase message delivery with reduced delay implying enhanced network reliability and low energy consumption. This chapter also makes the provision of multicast routing in an opportunistic network simulator to make it more realistic.

5.1 Literature Review

As discussed in Chap. 1 (Sect. 1.4.3), the routing protocols in an opportunistic network can be divided into two: forwarding-based (single-copy) routing and replication-based (multicopy) routing. The single-copy routing is rarely used in an opportunistic network in real-life scenarios due to its inherent limitations of marginal *delivery ratio* and unbound delivery latency. On the contrary, the multicopy routing protocols exhibit higher message delivery compared to their single copy counterpart due to the existence of numerous replicas of a single message. Such protocols replicate messages with the intention that at one replica reaches the destination. Both types of protocols can be further divided into two categories such as unicast and multicast routing protocols. Numerous unicast and multicast protocols are designed so far for

S. Bhattacharjee et al., *Post-disaster Navigation and Allied Services over Opportunistic Networks*, Smart Innovation, Systems and Technologies 228,
https://doi.org/10.1007/978-981-16-1240-4_5

routing in opportunistic networks. A few of the extensively used multicopy routing protocols are Epidemic [1], Spray and Wait [2], PRoPHET [3], PRoPHETv2 [4], MaxProp [5], RAPID [6], Bubble Rap [7], SnMPF [8], modified Epidemic routing [9], CAMR [10], RelayCast [11], SNAMD [12], SDM and MDM [13], multicast polymorphic epidemic routing [14], EASDM and EAMDM [15] etc. The protocols [1–9] are intended for unicast communication and [10–15] are intended for multicast communication. The routing protocols [7, 14, 15] are relatively recent reporting.

5.1.1 Unicast Routing Protocols

The idea of dissemination of epidemic diseases is employed in Epidemic routing protocol [1]. In this protocol, messages are forwarded from the creator node to all the encountered (relay) nodes in case the copy of the message is not present at the encountered nodes. The relay nodes similarly forward the messages. When the entire message buffer of a node becomes occupied, the oldest messages from the buffer are discarded. In this protocol, messages to be forwarded are chosen based on the FIFO strategy from the message buffer and a limit on the replication of messages is not imposed. Improving the possibility of message delivery and minimizing latency is a couple of key objectives of this protocol. Although the protocol presumes that the opportunistic network nodes exhibit arbitrary movement patterns, but in reality, nodes do not move in a purely arbitrary way. Hence, the protocol is not suitable for the majority of real-life scenarios. Moreover, this protocol employs the uncontrolled replication of messages to enhance the *delivery ratio*. Such uncontrolled replication increases network overhead substantially.

In another work [2], the limitation of Epidemic routing protocol is addressed and a controlled replication-based routing protocol Spray and Wait (SnW) is proposed. The spray and wait phase are the two primary phases of this protocol. A pre-defined quantity of copies of a specific message is sent to some unique relay nodes in the spray phase. Upon reception, the relay nodes further similarly send those messages. During the wait phase, the relay nodes accomplish the direct delivery of the messages in case the destinations are unreachable in the spray phase. When multiple nodes having identical message replicas arrive at the radio range of the destination node, a solitary replica which arrives first is retained by the destination node. All the other copies of that message are simply discarded.

In order to address the limitation of Epidemic routing regarding movement pattern, a Probabilistic Routing Protocol using History of Encounters and Transitivity (PRoPHET) is proposed in [3]. The PRoPHET employs a greedy algorithm and does not presume arbitrary movement patterns of nodes. The node to node messages forwarding takes place based on the assumption that the later node ensures a superior likelihood of meeting the destination than the former node. In PRoPHET protocol, the node mobility characteristics are predicted with the help of a metric called delivery predictability (DP). In each node, a DP table is kept for preserving the DPs with other nodes in the network. The DPs are calculated using the past encounters among the nodes and the transitive property. However, in [4] the authors discover some real-life scenarios where although some nodes exhibit extraordinary delivery predictabilities, the messages remain undelivered to the destinations owing to the absence of direct contact. The aforesaid shortcomings can be addressed by introducing an enhanced variety of the traditional PRoPHET protocol PRoPHETv2.

The protocols like Epidemic, SnW and PRoPHET do not offer buffer management functionality. Hence, the possibility of buffer overflow is present causing unsolicited message termination. The MaxProp routing protocol [5] is designed to eliminate such problems in opportunistic networks. In this routing protocol, a routing table is maintained at each node which describes the cost of reaching other nodes in the network through the existing neighbors. The messages are forwarded based on their cost of reaching the destination. This protocol employs history-based forwarding with an unlimited replication paradigm.

In another work [6], a resource allocation protocol for intentional DTN (RAPID) is suggested assuming routing to be a resource allocation problem in opportunistic networks. In RAPID the duplication of messages is controlled using a metric called per-message utility. In the case of buffer overflow, the messages with lower utilities are removed from the buffer. Just like MaxProp this protocol also follows history-based forwarding with an unlimited replication paradigm.

In order to make routing decisions, the social properties of the nodes are not considered in the aforesaid protocols [1–6]. The concept of social relations among the people is at first introduced in a routing protocol called Bubble Rap in [7]. This protocol uses a combination of a couple of social metrics such as centrality and community for forwarding messages. The nodes with high centrality values and associated with the identical community as a destination are chosen as relay nodes in this protocol. Apart from this, in some applications such as post-disaster scenarios, situational information dissemination requires priority-based message forwarding due to the existence of unimportant messages. Hence some of the unicast priority-based routing protocols are reviewed.

In one such work [8], a prioritized message routing protocol is proposed for opportunistic networks deployed in disaster areas by modifying Spray and Wait routing protocol. This protocol determines the priority of a message depending on the node's movement speed, the latest encounter time (time till the last encounter with the destination) as well as the distance of the node carrying the message from a destination.

In another work [9], three different message prioritization strategies such as Round Robin, Tiered and Oblivious are implemented on Epidemic routing protocol in an opportunistic network environment. The prioritization strategies mentioned in the work depending on age, forwarding history and the size of the messages.

Both the priority-based unicast routing protocols in [8, 9] does not consider message content while determining priority. But, in post-disaster scenarios, messages need to be prioritized based on content in order to filter out unimportant messages. Hence, the abovementioned routing protocols remain unsuitable for post-disaster scenarios.

All the aforementioned protocols are intended for unicast communication. However, some of the applications such as post-disaster management services (e.g. resource management, navigation) require periodic sharing of resource inventories and trajectory traces of volunteers among the designated group of volunteers from different disaster response agencies. Thus, unicast routing protocols remain unsuitable for resource management or navigation services. Hence, several multicast routing protocols are reviewed next.

5.1.2 Multicast Routing Protocols

Several multicast routing protocols are proposed in opportunistic networks. The works describing such protocols can be categorized into two, such as pure opportunistic and social-based [16].

5.1.2.1 Pure Opportunistic Approach

In one such work [10], a context-aware multicast routing (CAMR) protocol is proposed which relies on node densities in opportunistic networks. In this protocol, nodes identify their association with the sparse or the thick part of the network, by calculating the arithmetic mean of the quantity of neighbors encountered. If the nodes are associated with a higher number of neighbors than a threshold, then such nodes are associated with the thicker part of the network and vice versa. In the case of nodes positioned at more than two-hop distance, a multicast tree is constructed starting from source to all the destinations and messages are forwarded through the downward path. This routing protocol performs well under low node density (sparse network).

In another work [11], mobility-assisted multicast routing protocol (RelayCast) is proposed. RelayCast is an enhancement of the node mobility-based two-hop relay routing algorithm [17]. This protocol consists of two phases, such as relay and delivery phase. The messages are forwarded from the source to relay nodes or destination in the relay phase. A distinct queue is maintained for each destination at the relay nodes and the messages are forwarded from that queue to the destination in the delivery phase. Experimental results demonstrate the superiority of RelayCast in terms of throughput and scalability with low latency.

5.1.2.2 Social-Based Approach

In one of the works [12], a social network-aided multicast delivery (SNAMD) protocol is proposed. This protocol deals with node communities and connector nodes. The connector nodes can meet multiple communities. Precisely, this protocol disseminates a message replica to connector nodes and allows the nodes to utilize either the DP or the estimated node contact rate to forward messages.

In another work [13], single and multiple data multicasting protocols are studied in order to decrease the data dissemination cost by using the social characteristics of the nodes. Here, the dissemination cost is reduced by curtailing the quantity of relay nodes. The relay selection problem in this protocol is expressed as a unified knapsack problem. In the case of single data multicast (SDM), this protocol identifies relays with the help of the node's centrality. On the contrary, the relays are identified by using social community structures in the case of multiple data multicast (MDM).

In [14], a multicast polymorphic epidemic routing scheme is proposed for data forwarding in opportunistic networks. The likeness of the node's interest is the key to data forwarding in this work. In addition to that adaptive recovery strategies are also employed in this protocol to reduce network overhead.

In another work [15], energy-aware single and multiple data multicasting protocols are proposed. Here, the relay nodes are identified using the metrics centrality and residual energy. These protocols are the enhancement of [13] with respect to energy efficiency. Here, the network lifespan is enhanced by introducing a load balancing approach during the selection of relays. This approach decreases the overall energy consumption of the protocols.

To the best of our knowledge message priority-based multicast routing protocols for opportunistic networks have not been reported till date.

The routing protocols discussed in [12–15] support social-based multicast communication and in that way decrease network overhead. Though the competence regarding the low network overhead is attained at the cost of declining *delivery ratio* as well as the growing latency, hence the aforesaid multicast routing protocols [12–15] does not prove beneficial for the post-disaster scenarios.

5.2 Pen-PRoPHET: Priority-Enhanced PRoPHET Routing Protocol

Although the opportunistic network has been widely adopted for post-disaster communication, due to intermittent connectivity and constrained resources (mainly battery) of opportunistic network nodes, not all messages get forwarded in such a network. As a result, several crucial messages, waiting in the message buffer may get dropped and unimportant, extraneous messages (sentimental comments of victims or volunteers, etc.) may get transmitted at the cost of expensive network resources. Therefore, filtering out such irrelevant messages (containing unrelated information and emotional expressions) and prioritizing relevant messages according to their importance becomes crucial so that messages which are critical for situational awareness obtain the maximum priority and get disseminated in the network at the minimum possible time. In this section, the primary objective is to ensure best-effort delivery of emergency situational messages using a two-step approach: (i) segregation of high-priority messages through natural language processing-based filtering and (ii) dissemination of filtered messages over opportunistic networks using a Priority-enhanced PRoPHET (Pen-PRoPHET) routing protocol [18] which is developed by adapting the popular PRoPHET routing protocol [3, 4].

The proposed Pen-PRoPHET routing protocol segregates decisive messages and prioritizes them according to their significance. The protocol achieves content-based message segregation based on natural language processing (NLP) and a message prioritization based on the consequences of this segregation. The second level of on-the-fly prioritization is performed using the delivery predictabilities by adopting the PRoPHET routing protocol. The performance of the Pen-PRoPHET is assessed using the ONE simulator by comparing it with other well-known opportunistic network routing protocols.

5.2.1 Network Architecture

The proposed *three-tier* architecture discussed in Sect. 4.2.4 is employed as the principal architecture of the Pen-PRoPHET routing protocol. As mentioned in the proposed architecture the existence of three-node categories such as *volunteer node* ($node_{volunteer}$), *camp node* ($node_{camp}$) and *ferry node* ($node_{ferry}$) are primarily assumed in a post-disaster scenario. In addition to that, a *command and control node* ($node_{command}$) representing the command and control center is also incorporated. The functionalities of the aforesaid nodes in the network architecture are as follows:

- *Volunteer node* ($node_{volunteer}$): Volunteers equipped with smartphones are considered as *volunteer nodes*. The *volunteer nodes* collect situational information intended for *camp nodes*. In addition to that, they send such information to other *volunteer nodes*.
- *Camp node* ($node_{camp}$): Camps consist of Throwboxes that are treated as *camp nodes*. Such nodes accumulate the situational information from the *volunteer nodes*, collate them and generate comprehensive messages containing overall situational information destined for the other *camp nodes* or the *command and control node*.
- *Ferry node* ($node_{ferry}$): Supply vehicles equipped with smartphones or tablets are considered as *ferry nodes*. They typically carry situational messages intended for remote *camp nodes* and *command and control nodes*. *Ferry nodes* facilitate long-distance communication in post-disaster scenarios.
- *Command and control node* ($node_{command}$): The command and control station is equipped with one or more *command and control nodes* which are high-end desktops computers or servers designed to accumulate situational information from various *camp nodes* and other sources. After taking necessary action depending on situational information such nodes generate corresponding messages (e.g. resource allocation messages, resource mobilization messages, instructions, etc.) destined for *camp nodes* or *volunteer nodes*.

Thus, post-disaster message dissemination involves the communication of

- Situational messages from $node_{volunteer}$ to $node_{camp}$, $node_{camp}$ to $node_{ferry}$ and from $node_{ferry}$ to $node_{command}$.
- Resource allocation, mobilization messages or instructions from $node_{command}$ to $node_{ferry}$, $node_{ferry}$ to $node_{camp}$s and $node_{camp}$s to $node_{volunteer}$s.
- Conversation and sentimental messages from $node_{volunteer}$ to $node_{volunteer}$.

5.2.2 *Categorization of Messages in Post-disaster Scenarios*

Real WhatsApp chats exchanged between DFY team members during post-disaster relief operations (April 2015 to July 2015) in Nepal are used as datasets. Figure 5.1 depicts a sample WhatsApp conversation among the team members of a pan-India humanitarian organization Doctors for You (DFY) during the 2015 Nepal disaster. These messages are typically unstructured and written in English and other regional languages. Among those messages, we consider only the ones written in English. In order to categorize messages generated during post-disaster response operations

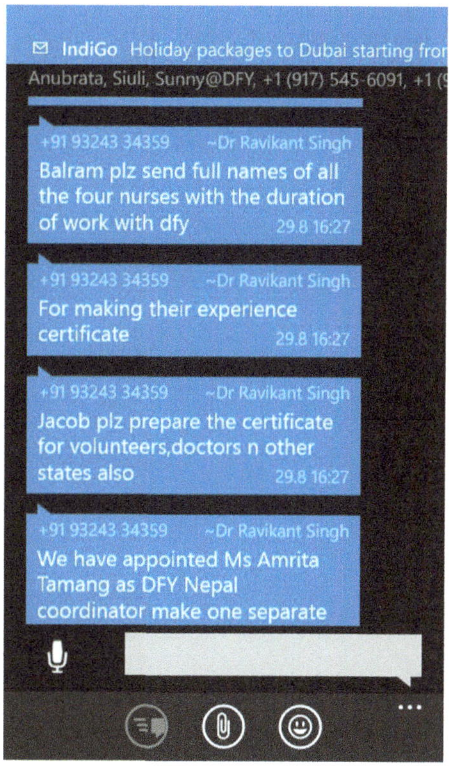

Fig. 5.1 Sample WhatsApp conversation between DFY team members

based on message content, we analyze the WhatsApp chats generated by the volunteers of DFY. The WhatsApp chats are not usually stored on servers. Therefore, the WhatsApp chats stored in smartphones are sent to the server as email attachments in order to get access to the WhatsApp data. The conversations are analyzed manually. The analysis of the messages reveals that the messages generated can be broadly distributed in five categories based on the content, viz., resource requirement, resource allocation, situational, conversation and sentimental. Table 5.1 demonstrates the message categories illustrated with real examples.

Based on such categorization, the messages are prioritized as per the assumptions mentioned below. We assume that resource requirement messages are of utmost importance in a post-disaster scenario. However, the majority of these messages represent critical resource requirements to sustain the basic survival needs.

The next higher priority is allocated to the situational messages which are important to the disaster response agencies to analyze the situation and take appropriate action. Resource allocation messages are prioritized after situational messages followed by conversation and sentimental messages. Table 5.2 presents the message priority table showing different message categories on a scale of five.

Table 5.1 Message categorization table for WhatsApp messages in Nepal disaster

Message category	WhatsApp messages from Nepal disaster
Resource requirement	We need gynecology doc for two weeks
	Drug list Vitamin C 1 lakh, Calcium tablets 1 lakh, inhaler 500
	We need one month supply
Resource allocation	Digital Xray machine dispatched to Bidur
	15 tents are distributed in Nuwakot
	Team going to madankot
Situational	7.1 earthquake in Nepal
	Same here..not felt.
	All safe
Conversation	Ranjeet reached Nepal or not…? Please confirm
	Yes sir he reached yesterday
	Where are you? Shandeepan
Sentimental	Keep up the good work dfy
	Thanks Anurag and Ravi!
	Ravikant Sir… Remembered my Days at Panchkanya

Table 5.2 Message priority table

Message category	Message priority (out of 5)
Resource requirement	5
Situational	4
Resource allocation	3
Conversation	2
Sentimental	1

Since content-based categorization of unstructured messages requires a fair amount of computational intelligence and intervention of machine learning techniques, categorization and prioritization of a large number of unstructured messages based on a keyword search are not feasible. So, Naïve Bayesian Classifier (NBC) is employed in order to categorize messages based on content. NBC performs well in classifying messages with lower training data and also exhibits superior resilience to missing data compared to other classification techniques [19]. Moreover, it is easy to implement NBC modules and the modules perform well with energy constraints. These benefits encourage us to use NBC to classify messages in an opportunistic network environment.

5.2.3 Limitation of PRoPHET Routing Protocol

This section illustrates the limitations of the existing PRoPHET protocol [3, 4] in post-disaster scenarios.

In this protocol, a DP table is retained by each node. The DPs are typically calculated using the past encounters among the nodes and the transitive property. When nodes encounter the DP tables are exchanged among themselves. In addition to that, the nodes modernize their individual DP table as per the following equation:

$$P(A, B)_{new} = P(A, B)_{old} + (1 - P(A, B)_{old}) \times P_{enc} \tag{5.1}$$

Here, P_{enc} is a configurable parameter which makes sure that the frequently encountered nodes acquire maximum DP.

As a result of the transitive property of the DP, A updates DPs for all the destinations i recognized by B depending on the DP table stored at B. Transitive update is usually performed with the help of Eq. (5.2) where β is a configurable parameter:

$$P(A, i)_{new} = \max(P(A, i)_{old}, P(B, i) \times P(A, B)_{new} \times \beta) \tag{5.2}$$

The eradication of outdated data from the DP table is achieved by gradually aging the entries of the table as per Eq. (5.3). As per the equation, γ is a constant and T is the amount of time till the earlier aging was executed.

$$P(A, i)_{\text{new}} = P(A, i)_{\text{old}} \times \gamma^T \tag{5.3}$$

The PRoPHET routing protocol inherently supports unicast communication of messages depending on the past encounters among the nodes. The message priority is not considered while forwarding messages in the PRoPHET routing protocol. As a result, crucial messages might be delayed or discarded. This limitation restricts the application of the original PRoPHET protocol in post-disaster situational awareness service.

5.2.4 Proposed Pen-PRoPHET Communication Architecture

In this section, we propose a system to ensure best-effort delivery of emergency circumstantial messages using a two-step approach: (i) segregation of high priority messages through NLP based filtering and (ii) dissemination of filtered messages over opportunistic networks using a Priority-enhanced PRoPHET (Pen-ProPHET) routing protocol. The communication architecture of the proposed Pen-PRoPHET routing protocol is illustrated in Fig. 5.2. The proposed communication architecture consists of three modules, viz., classification module, prioritization module and Pen-PRoPHET module and a priority queue.

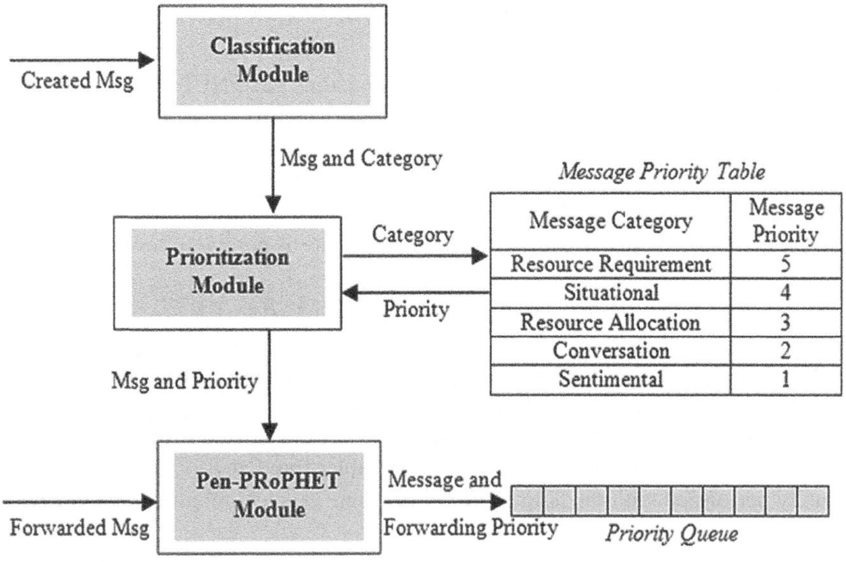

Fig. 5.2 Communication architecture of the proposed Pen-PRoPHET routing protocol

5.2.4.1 Classification Module

This module consists of a Naïve Bayesian Classifier (NBC), which classifies a newly generated message in one of the five categories mentioned in Table 5.1. Here, NBC is used for message categorization as it exhibits superior support toward lost data than the SVM classifier [19]. Moreover, it is simple to implement and the data required to train NBC is less compared to the other models.

The Weka Version 3.6 (machine learning tool) [20] is used for message classification. The Weka is chosen as it is open source and includes 76 different classification/regression algorithms. The NBC is trained using 200 text messages (40 messages from each category) received from WhatsApp conversations among DFY volunteers.

The classification module is associated with every node connected to the network and becomes active after new message creation. This module receives newly created messages as input from each node and classifies them into pre-defined categories according to the message content. The message along with the attached category information is generated as output from this module. Subsequently, the output is directed to the prioritization module for processing.

5.2.4.2 Prioritization Module

This module receives messages along with its category as input from the classification module. It consults the category with a lookup table called message priority table and assigns a positive integer value called message priority (P_m) on a five-point scale. The message along with the attached P_m is generated as output from this module and forwarded to the Pen-PRoPHET module.

5.2.4.3 Pen-PRoPHET Module

This is a message forwarder module based on the PRoPHET routing algorithm. This module computes the *forwarding priority* (P_f) of every message received from the prioritization module or a forwarding node, based on message priority and DP of the forwarder node. *Forwarding priority* is an essential component for on-the-fly message prioritization.

In traditional PRoPHET routing, a forwarder node is identified if the DP (as stated in Sect. 5.2.3) of the destination is greater at that node. This technique substantially reduces message duplication and reduces network overhead since messages are forwarded only to nodes with a greater likelihood of meeting the destination. In order to incorporate the aforesaid strength of the PRoPHET routing algorithm, a metric called *forwarding priority* ($P_f \in [0, 1]$) is introduced. P_f is a weighted sum of P_m and DP (P_d)

$$P_f = W_1 \times \frac{P_m}{5} + W_2 \times P_d$$

W_1, W_2 are priority weights, where $W_1 + W_2 = 1$. P_m is represented on a five-point scale. In order to scale down its value between 0 and 1, it is divided by five.

The introduction of P_f improves the forwarding mechanism of prioritized messages by controlling the replication of messages in the network. It also reduces the possibility of starvation of the comparatively lower priority messages. For instance, a node N_1 tries to send two messages M_1 and M_2 with P_m values 5 and 4, respectively, to the nodes N_3 and N_4. Later it encounters node N_2. If the P_d of nodes N_3 and N_4 are 0.5 and 0.8, respectively, at N_2, then P_f of M_1 becomes 0.75 and M_2 becomes 0.8. Thus, the message with relatively inferior priority (M_2) receives the opportunity of forwarding.

The Pen-PRoPHET module can be operated with three different settings.

- ***Setting 1***: When the values of W_1 and W_2 are both equal to 0.5 messages are forwarded based on both P_m and P_d.
- ***Setting 2***: When the value of W_1 is 0 and W_2 is 1, messages are forwarded based on P_d only like the traditional PRoPHET routing algorithm.
- ***Setting 3***: When the value of W_1 is 1 and W_2 is 0 messages are forwarded based on P_m.

The message along with the P_f as output is dispatched and stored in the appropriate position of the priority queue (*PQ*) and when the opportunity arrives message will be sent to the suitable node according to its P_f.

An algorithm explaining the workflow of the classification, prioritization and Pen-PRoPHET modules is presented below. The algorithm runs on the smartphones or PDAs used by volunteers and relief workers, enabling them to do in-network processing of generated and received messages. The messages without *forwarding priority* are the inputs to the algorithm. On the contrary, the messages with *forwarding priority* after prioritization are the outputs from the algorithm.

Algorithm: Classification and Prioritization
Input: Message (*msg*)
Output: Message (*msg*) along with Forwarding Priority (P_f)

```
Begin
do
    if msg_source_id == node_id then        /* when new messages arrive */
        category = categorize (msg)
        P_m = prioritize (msg, category)
        P_f = W_1 × P_m/5 + W_2 × P_d
    else
        P_f = W_1 × P_m/5 + W_2 × P_d
    end if
        PQ = P_f                            /* P_f is stored in the priority queue (PQ) */
while N_f ≠ NULL                            /* N_f is set of all forwarder nodes */
End
```

5.2.5 *Performance Evaluation*

The performance of the proposed Pen-PRoPHET protocol is evaluated in this section through comparison with well-known opportunistic network routing protocols like PRoPHET, Spray-And-Wait, MaxProp and Epidemic with respect to *delivery ratio* and *overhead ratio.*

5.2.5.1 Simulation Setup

The experiments are conducted using ONE simulator, version 1.5.1-RC2 [21], based on the Nepal disaster map, presented in Fig. 3.9 of Chap. 3. The simulation environment is identical to the environment mentioned in Chap. 3 (Sect. 3.3.4.1). The number of *volunteer nodes* for simulation is set to 90, based on the proposed mobile node deployment plan mentioned in Chap. 4 (Sect. 4.3.2). Two supply vehicles (*ferry nodes*) repeatedly visit the camps from the command and control center. It is assumed that messages are generated every 1800–2100 s. The PDM model is employed as the underlying mobility model for *volunteer nodes*. The supply vehicles follow the map route movement model. The *three-tier* architecture mentioned in Chap. 4 (Sect. 4.2.4) is chosen for the simulation. During the simulation, *volunteer nodes* are treated as *tier-1* devices, *camp nodes* as *tier-2* devices and *ferry nodes* as *tier-3* devices. The general parameters of the simulation are represented in Table 5.3.

Table 5.3 Parameter values used in the simulation of Pen-PRoPHET routing protocol

Parameters	Values
Simulation time	43200 s (12 h)
Number of nodes	96 (volunteer: 90, Camp: 4, Ferry: 2)
Generation rate of messages	One new message per node every 1800–2100 s
Size of messages	0.5–1 MB
Buffer size	5 MB
Message TTL	100 min
Routing protocols	MaxProp Spray and Wait Epidemic PRoPHET Pen-PRoPHET

5.2.5.2 Simulation Metrics

The *delivery ratio* and *overhead ratio* are chosen as metrics for analyzing simulation results. Both the metrics are already defined in Chap. 4 (Sect. 4.2.5.2) and Chap. 3 (Sect. 3.3.4.2), respectively.

5.2.5.3 Results and Discussion

This section provides the analysis of simulation results collected from varying simulation scenarios. To assess the performance of the Pen-PRoPHET, four sets of experiments are conducted by running simulation for 12 h. In the first set of experiments, the distribution of *delivery ratio* based on the message priority in Pen-PRoPHET is assessed. Figure 5.3a shows the results. The results clearly indicate that the *delivery ratio* increases with growing message priority. It is further observed from the results that the priority-5 messages demonstrate around 57.14% improvement in terms of *delivery ratio* compared to priority-1 messages. This implies the messages with higher priority are mostly delivered, whereas messages with lower priority are dropped.

In the second set of experiments, Fig. 5.3b provides the comparative results of *delivery ratio* for all the competing routing protocols including Pen-PRoPHET operated with Setting 1. The experimental results reveal that the Pen-PRoPHET protocol operated with Setting 1 exhibit around 14.28%, 9.6%, 23.88% and 19.08% improvement in terms of *delivery ratio* compared to PRoPHET, MaxProp, Spray and Wait and Epidemic protocols, respectively. Thus, the results reveal the superiority of Pen-PRoPHET routing while delivering the highest priority messages compared to other conventional opportunistic network routing protocols. It is further observed that although the MaxProp routing protocol exhibits a superior overall *delivery ratio*, it

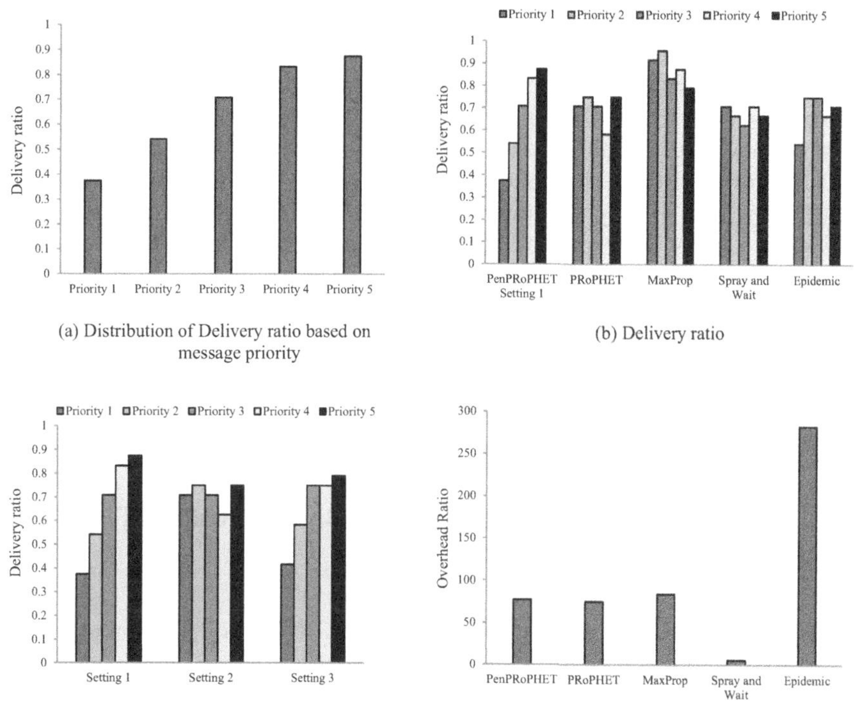

(a) Distribution of Delivery ratio based on message priority

(b) Delivery ratio

(c)Three different settings of Pen-PRoPHET

(d) Overhead ratio

Fig. 5.3 Comparative performance of Pen-PRoPHET routing protocol

falls behind while the *delivery ratio* of the highest priority messages is taken into consideration. The results clearly reveal that Pen-PRoPHET routing outperforms PRoPHET routing while delivering higher priority messages.

In the third set of experiments, Fig. 5.3c demonstrates the comparison of three different settings of Pen-PRoPHET routing as mentioned in Sect. 5.2.4.3 with respect to the *delivery ratio*. The results comply with our assumptions. The results demonstrate that for Setting 1 ($W_1 = 0.5$ and $W_2 = 0.5$) higher priority messages outperform lower priority messages in terms of *delivery ratio*, which a typical Pen-PRoPHET routing characteristic. Setting 2 ($W_1 = 0$ and $W_2 = 1$) exhibits the characteristics of traditional PRoPHET routing where messages are forwarded depending on P_ds only and priority is not considered. As a result, messages are delivered arbitrarily irrespective of priority, which is clearly evident from simulation results. Setting 3 ($W_1 = 1$ and $W_2 = 0$) forwards messages based on P_ms only, and P_ds are not considered. Thus, the overall *delivery ratio* is reduced.

In the fourth set of experiments, Fig. 5.3d provides the comparative results of the *overhead ratio* for all the competing routing protocols including Pen-PRoPHET. The experimental results demonstrate that the Spray and Wait routing protocol exhibits the lowermost *overhead ratio* among the conventional opportunistic network routing protocols followed by the Pen-PRoPHET routing protocol. On the contrary, the Spray and Wait protocol achieves this goal at the cost of a reduced *delivery ratio* which is undesirable in post-disaster scenarios. Thus, having said the intrinsic boundaries of the Pen-PRoPHET routing, it shows merit in delivering critical messages to the destination with reasonable message delivery cost in terms of replication.

The Pen-PRoPHET protocol facilitates the dissemination of situational messages based on message priority. However, the messages are forwarded to one destination only. Alternatively, the Pen-PRoPHET routing protocol is incapable of forwarding messages in multicast mode. Thus, this protocol does not prove suitable for post-disaster management services which may require to deliver at multiple destinations. To be more specific, the resource management service requires multicasting of resource inventories and queries. Hence, in the next section, suitable routing protocols supporting post-disaster resource management service is proposed.

5.3 Ge-PRoPHET: Group Encounter-Based PRoPHET Routing Protocol

In this section, the Group encounter-based PRoPHET (Ge-PRoPHET) protocol [22] is discussed in order to facilitate resource management service over opportunistic networks in post-disaster scenarios. As discussed in Chap. 3, one of the crucial parts of the disaster response operation is resource management. Camps set up by different disaster response agencies manage their own resource inventories in a decentralized manner. The resource inventories are typically produced by the camps. The *camp nodes* broadcast those inventories periodically amid the *volunteer nodes* connected with the designated response groups. Each resource inventories have a stipulated lifetime after which the resource inventory becomes outdated. Thus, group-based multicasting (*groupcasting*) of resource inventories within a specified time is utmost crucial. Likewise, recognizing a node that will respond to a query is almost impossible in advance. Hence, *groupcasting* becomes vital in query dissemination. Therefore, the original PRoPHET protocol is modified to enable *groupcasting*.

5.3.1 Rationale Behind Adapting PRoPHET

During the disaster aftermath, the volunteers associated with specific response groups visit the shelters regularly for supplying aids. A *volunteer node* associated with a specific response group meets *volunteer nodes* of a similar group or related groups frequently compared to the nodes of other response groups. Hence, it is beneficial to utilize such meeting patterns for the dissemination of resource inventories and queries. Thus, PRoPHET proves suitable for such previous meeting-oriented forwarding of resource inventories and queries.

Although some well-known opportunistic network routing protocols such as MaxProp [5] and BubbleRap [7] demonstrates superiority with respect to message delivery compared to PRoPHET, the protocols prove unsuitable for resource management service in post-disaster scenarios. In the following, the primary shortcomings of MaxProp and BubbleRap protocols are discussed.

The MaxProp typically incorporates an acknowledgment mechanism to eliminate the redundant copies of the delivered messages from the network. Upon reaching the destination, a message typically generates the acknowledgment. The other nodes simply discard the copies of the delivered message retained by them from the message buffers in order to evade buffer overflow and decrease network overhead. On the contrary, resource management service necessitates the delivery of a solitary message to a response group (multicasting). The acknowledgment mechanism employed by MaxProp will hinder multicasting by eliminating all the replicas of a message from the network when at least one copy arrives at the destination. Hence, the MaxProp routing protocol proves unsuitable for post-disaster resource management service.

BubbleRap protocol utilizes the social characteristics such as centrality and community of the nodes to disseminate messages. A couple of centralities metrics such as global and local are employed by this protocol. Here, inter- and intra-community encounters among the nodes are signified by the global and local centrality, respectively. Initially while forwarding messages, the relay nodes are identified depending on greater global centrality values. Upon reaching the identical community node, the message is disseminated to other nodes associated with that community using local centrality. This protocol has several limitations. One such limitation is that the high centrality of relay nodes rarely ensures a superior likelihood of meeting the destinations. In addition to that, the community detection algorithm employed by the protocol fails to detect the appropriate community of a relay node in approximately 15% of cases. Moreover, due to the existence of overlapping community structure, the protocol does not produce a reasonable performance in post-disaster scenarios. Thus, BubbleRap is unsuitable for post-disaster resource management service.

The aforementioned shortcomings of MaxProp and BubbleRap protocols are the primary inspiration behind modifying the PRoPHET protocol to facilitate resource management service in post-disaster scenarios.

5.3.2 Proposed Ge-PRoPHET Routing Protocol

The discussion in Sect. 5.2 indicates the necessity of *groupcasting* in post-disaster resource management. Consequently, the Group encounter-based PRoPHET (Ge-PRoPHET) is designed. It is an enhancement of the well-known PRoPHET protocol to enable *groupcasting* of resource inventories and queries. Unlike PRoPHET, a group delivery predictability (GDP) table is preserved at each node in Ge-PRoPHET. The DP of each node in the network, with all response groups available in the affected area, are preserved in the GDP table. The GDP of a *volunteer node X*, with any node of a group *G*, is presented as *P(X, G)*. When two *volunteer nodes* *X* and *Y* are associated with similar or dissimilar response groups meet, the exchange of GDP tables takes place among the nodes. In addition to that, both nodes update their GDP tables as per Eqs. (5.4) and (5.5)

$$P(X, G) = P(X, G)_{\text{old}} + (1 - P(X, G)_{\text{old}}) \times P_{\text{enc}} \tag{5.4}$$

$$P(Y, G) = P(Y, G)_{\text{old}} + (1 - P(Y, G)_{\text{old}}) \times P_{\text{enc}} \tag{5.5}$$

Here, *P*(*X*, *G*) and *P*(*Y*, *G*) are dissimilar since *X* might not encounter as many *volunteer nodes* associated with *G*, as *Y* and vice versa. P_{enc} is a configurable parameter. As per the PRoPHET, the suggested value of P_{enc} is 0.75 [23].

Owing to the transitive property of the GDP, *X* and *Y* modify their GDP tables for entire groups (*k*) with the help of Eqs. (5.6) and (5.7) where β is a configurable parameter. The suggested value of β as per the PRoPHET is 0.25 [23].

$$P(X, G_k)_{\text{new}} = \max(P(X, G_k)_{\text{old}}, P(Y, G_k) \times P(X, G_k)_{\text{new}} \times \beta) \tag{5.6}$$

$$P(Y, G_k)_{\text{new}} = \max(P(Y, G_k)_{\text{old}}, P(X, G_k) \times P(Y, G_k)_{\text{new}} \times \beta) \tag{5.7}$$

If the encounter between nodes is infrequent then the GDP value needs to be reduced due to aging. The aging in the GDP table is performed using Eq. (5.8) where γ is a constant. The suggested value of γ as per the traditional PRoPHET is 0.998 [23].

$$P(X, G_k)_{\text{new}} = P(X, G_k)_{\text{old}} \times \gamma^T \tag{5.8}$$

Due to the replication-based nature of Ge-PRoPHET, adjusting unintentional duplication of resource inventories or queries becomes crucial for decreasing the network overhead and energy consumption. Here, a node is selected for forwarding resource inventory/query, if the GDP of the destination group of the resource inventory/query is better at the contacted node. The unintentional duplication of resource inventories/queries is reduced by this technique as resource inventories/queries are disseminated to the nodes having a higher likelihood of meeting the destination group and consequently the energy consumption due to transmission is also reduced.

5.3.3 Performance Evaluation

This section provides a performance assessment of the Ge-PRoPHET protocol through the simulation of the inventory sharing module of DPDRM as discussed in Chap. 3 (Sect. 3.3.2.1). The results are also compared with three existing unicast protocols, such as PRoPHET, Epidemic, MaxProp, and two multicast protocols

Table 5.4 Parameter values used in the simulation of Ge-PRoPHET routing protocol

Parameters	Value
Simulation time	2 h
Number of nodes	99 (volunteer: 90, Camp: 7, Ferry: 2)
Transmission range	80 meters
Buffer size	20 MB
Message TTL	720 min
Message generation rate	One in every 12 h (resource inventory)
Mode of communication	Group-based multicasting (*groupcasting*)

such as SDM (single data multicast) [13] and EASDM (energy-aware single data multicast) [15].

5.3.3.1 Simulation Setup

Here, the simulation environment is configured in the modified ONE simulator [24] which facilitates a multicast mode of communication. The simulation setup is similar to the setup as mentioned in Chap. 3 (Sect. 3.3.4.1). On the map of the affected area of Kathmandu, as demonstrated in Fig. 3.9 of Chap. 3, seven camps are set up within an area of 3.5 km^2 around the Patan area. The other parameters such as the number of *volunteer nodes*, *ferry nodes*, mobility models and network architecture are similar to Sect. 5.2.5.1. However, unlike Sect. 5.2.5.1, here the number of camp nodes is increased to seven. Moreover, as opposed to Sect. 5.2.5.1, in this section, the mode of communication for simulation is chosen as group-based multicasting.

It is assumed that the resource inventories are *groupcast* from *camp nodes* to the other nodes associated with similar groups periodically, say once a day. As mentioned in Chap. 4 (Sect. 4.2.5.1), the disaster response operations are typically carried out for 12 h a day from 6 a.m. to 6 p.m. Hence, the update and sharing of resource inventories take place at 12 h interval. Thus, Message TTL is set to 12 h (720 min). Other important simulation parameters are given in Table 5.4.

5.3.3.2 Simulation Metrics

The performance of the Ge-PRoPHET is measured by two types of parameters such as design parameters and network parameters. The design parameters include *inventory share rate*, *overhead ratio* and *energy consumption* whereas *latency* is considered as a network parameter.

Inventory share rate: It measures the level of resource inventory dissemination among the nodes of the network within a fixed period of time. The *inventory share rate* is expressed as follows:

$$\text{Inventory share rate (\%)} = \frac{\text{Number of nodes received resource inventories till a particular instant of time}}{\text{Total number of nodes in the network}} \times 100\%$$

The other three metrics such as *overhead ratio*, *average delivery latency* and *average energy consumption* are already defined in Chap. 3 (Sect. 3.3.4.2).

5.3.3.3 Results and Discussion

Four sets of experiments are conducted for measuring performance in terms of both the design and network parameters. The effectiveness of using various routing protocols in group-based resource inventory sharing is estimated through the first set of experiments. Figure 5.4a shows the results of this set of experiments. We primarily observe that for all the routing protocols, the *inventory share rate* rises steadily over time. To be more specific, after the entire simulation period, the *inventory share rate*

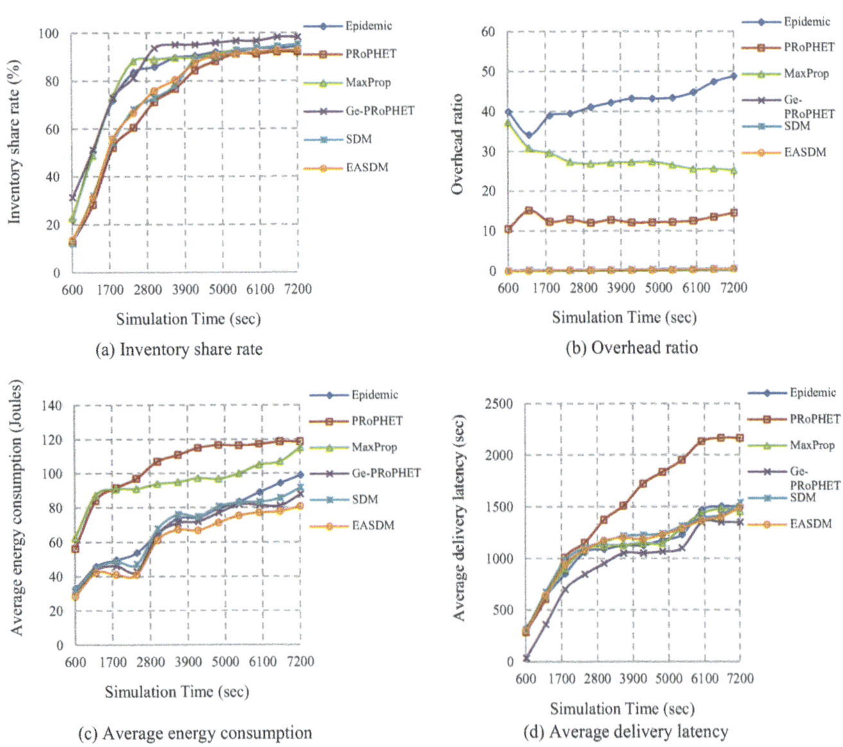

Fig. 5.4 Comparative performance of Ge-PRoPHET in inventory sharing

is 93.07% for EASDM, whereas this value is 95.38%, 98.46%, 95.38%, 92.3% and 94.61% for SDM, Ge-PRoPHET, MaxProp, PRoPHET and Epidemic, respectively. The results imply that the Ge-PRoPHET disseminates 98.46% resource inventories in 2 h from the generation of such inventories. This percentage value is the highest among all the competing routing protocols establishing Ge-PRoPHET's superiority over the other protocols.

The overhead of using various routing protocols is estimated through the second set of experiments. Figure 5.4b shows the results of this set of experiments. We observe that the *overhead ratio* rises marginally over time in all the cases. We also observe that, at the end of the entire simulation period, it is 0.76 for Ge-PRoPHET, whereas this value is 0.73, 0.79, 25.32, 14.67 and 48.96 for EASDM, SDM, MaxProp, PRoPHET, and Epidemic, respectively. It is clear from the plot that the network overhead implying message replication is insignificant in Ge-PRoPHET, EASDM and SDM. To be more specific, these three protocols incur 94.8%, 95% and 94.6% less overhead than PRoPHET. As the other three protocols such as MaxProp, PRoPHET and Epidemic are unicast in nature, they implement multicast by generating multiple copies of the message with varying message ids and disseminating all these messages in unicast mode. This results in a higher *overhead ratio* for these three protocols.

The *average energy consumption* of each node over time is estimated through the third set of experiments. Figure 5.4c plots the results of this set of experiments. We observe from the results that for all the competing protocols, including Ge-PRoPHET, the energy consumption increases over time. Precisely, at the end of the entire simulation period, Ge-PRoPHET consumes an average 88.01 J, whereas EASDM, SDM, MaxProp, PRoPHET and Epidemic consume 81.2 J, 91.96 J, 115.35 J, 118.93 J and 99.17 J, respectively. The results imply that EASDM and Ge-PRoPHET are relatively more energy saving compared to the others.

The *average delivery latency* over time is estimated through the fourth set of experiments and the results are shown in Fig. 5.4d. We observe from the results that the *average delivery latency* rises over time for all the competing protocols including Ge-PRoPHET. We also observe that, at the end of the entire simulation period, it is 1355.73 s for Ge-PRoPHET, whereas this value is 1497.23 s, 1545.23 s, 1461.12 s, 2171.91 s and 1508.57 s for EASDM, SDM, MaxProp, PRoPHET and Epidemic, respectively. The results clearly signify the superiority of Ge-PRoPHET over the other protocols with respect to the network parameter *average delivery latency*.

Summarily, the Ge-PRoPHET shows its dominance primarily with respect to *inventory share rate* which is one of our designated design goals. It also shows its superiority in maintaining network parameter value *average delivery latency*. Regarding the other two design goals *average energy consumption* and *overhead ratio*, the Ge-PRoPHET exhibits better performance overall than the other protocols, except for EASDM which again lacks in performing in terms of *inventory share rate* and *average delivery* latency. Thus, the Ge-PRoPHET shows its appropriateness in implementing the inventory sharing module of DPDRM.

Routing protocols discussed in Sects. 5.2 and 5.3 are based on the PRoPHET routing protocol which employs controlled replication of messages. Controlled replication obstructs the broadcasting of messages which is crucial for sharing trajectory traces accumulated by a node with all the other nodes in a network. Hence, both the routing protocols mentioned in the earlier sections cannot be employed for offering map-based navigation services. Thus, a novel routing protocol is discussed in the following section which facilitates offering map-based navigation services in post-disaster scenarios.

5.4 Trace Route Routing Protocol

In order to create the digital route map of an affected area, the GPS traces accumulated by the *volunteer nodes* need to be shared with other *volunteer nodes* and *camp nodes* positioned in that area. The sharing of trajectory traces is usually achieved through a group-based publish-subscribe technique [25]. As discussed in Chap. 2 (Sect. 2.3.1), in the disaster aftermath the *volunteer nodes* and other nodes (volunteer or camp) positioned in the identical area are the publisher and subscriber of GPS (trajectory) traces, respectively. Like this, the traces accumulation from an affected area takes place progressively at *volunteer nodes* and *camp nodes*. The route map of an affected area is constructed by assembling such traces. The group-based multicasting (*groupcasting*) is indispensable for the dissemination of trajectory traces among the nodes with reduced network overhead. Hence, the dissemination of traces is performed using *groupcasting*.

5.4.1 Rationale Behind Adapting Epidemic Routing Protocol

The dissemination of trajectory traces using *groupcasting* resembles the epidemic spreading. The Epidemic routing protocol [1] proves suitable for such scenarios. The protocol is designed using the notion of spreading epidemic infections in reality. Here, the replicated messages are forwarded to each encountered node with an intention of improving the prospects of message delivery and decreasing latency. However, in opportunistic networks, the Epidemic protocol is mainly used for unicast communication causing excessive network overhead and energy consumption as opposed to the prevailing multicast protocols [13–15]. Hence, the conventional Epidemic protocol remains unsuitable for disseminating GPS traces in the disaster aftermath.

5.4.2 Proposed Trace Route Routing Protocol

In order to address the aforesaid limitations, the original Epidemic protocol is modified and Trace Route protocol [26] is designed to enable *groupcasting* of GPS traces. The traces gathered by a node till an interval can be distinctively recognized by $Trace\_t_{creator\_id}$ where *creator_id* resembles the distinctive identifier of the creator node and t is the timestamp of trajectory traces gathering. Each $Trace\_t_{creator\_id}$ contains a collection of tracepoints (p). Each p consists of 3-tuple $p = (lat, long, t)$ where *lat* and *long* represent the latitude and longitude of the point collected at timestamp t, respectively. As opposed to Epidemic protocol, every trajectory trace in Trace Route is disseminated to a pre-defined group of destinations (*destination group*) depending on the location of the originator node. For instance, *trajectory traces* accumulated by a *volunteer node* will be forwarded to the other nodes (volunteer, camp and ferry) positioned in the same area. Thus, a *destination group* is formed. Each node keeps a list of the trajectory traces ($Trace\_list_{node\_id}$) stored at each node, where *node_id* is the distinctive identifier of the node. The $Trace\_list_{node\_id}$ contains a set $\{Trace\_t_{creator\_id}: Trace\_t_{creator\_id} \in Trace\_list_{node\_id}\}$. When a couple of nodes (X and Y) meet, their respective $Trace\_list_{node\_id}$ is exchanged. If X (source node) wants to send trajectory traces to Y (encountered node) and vice versa, the set difference operation is executed on the trace lists at both the nodes as follows:

$$Trace\_list_{source} - Trace\_list_{encountered} = \{Trace\_t_{creator\_id} \mid Trace\_t_{creator\_id} \in Trace\_list_{source} \text{ and } Trace\_t_{creator\_id} \notin Trace\_list_{encountered}\} \tag{5.9}$$

After this operation, nodes are able to identify their required trajectory traces. Subsequently, the *destination groups* of the designated trajectory traces are tested against the encountered *node_id*. The trajectory traces whose *destination group* includes the encountered *node_id* are sent to the trace queue for forwarding. This technique decreases the communication overhead significantly causing a reduction in energy consumption of the entire network. The communication overhead is reduced at the cost of computation overhead. Thus, the prospect of unintentional duplication of trajectory traces is eradicated in Trace Route.

5.4.3 Performance Evaluation

In this section, the performance of the Trace Route routing protocol is estimated. The performance is also compared with some important unicast protocols such as MaxProp, PRoPHET, Epidemic and two multicast protocols such as SDM [13] and EASDM [15].

Table 5.5 Power consumption readings of Wi-Fi in Samsung J7 Prime

Wi-Fi state	Power consumed (mW)	Energy consumption rate (J/s)
Wi-Fi off	0	0
Wi-Fi idle	164.79	0.165
Wi-Fi scanning	976.71	0.976
Wi-Fi receiving	1717.71	1.717
Wi-Fi sending	1786.24	1.786

5.4.3.1 Simulation Setup

The performance evaluation of Trace Route is done through simulation using ONE simulator. The simulation setup is identical to the setup discussed in Sect. 5.3.3.1.

We consider Samsung Galaxy J7 Prime smartphones with varying capacities as the devices are carried by different nodes such as volunteers, ferry nodes and *camp nodes*. For example, the *camp nodes* would carry such smartphones with higher secondary storage. The other specifications of the devices are Android v7.0 (Nougat), 3300 mAh, 3.85 V Li-ion battery.

Similar to the energy measurement methodology described in Chap. 4 (Sect. 4.2.5.3), we measure the communication (Wi-Fi) energy of such devices using the Trepn profiler [27] version 6.2. Also, following a similar method in Sect. 4.2.5.3, we get the initial stored energy as 45738 joules. Now, to configure the energy module of ONE simulator, in Table 5.5 power consumption-related values of the specified smartphones are provided for all the Wi-Fi communication activities. With the help of such values, the average energy consumption of the nodes for running the map builder module using the Trace Route routing protocol is to be measured.

5.4.3.2 Simulation Metrics

To estimate the performance of the *Trace Route Protocol*, we consider four simulation metrics such as *trace share ratio, overhead ratio, average energy consumption* and *delivery latency*. Out of these four metrics, we consider the first three as design metrics which have an impact on the major design issues. The rest of these metrics such as the *delivery latency* is considered as network performance metrics. All the metrics except the *trace share ratio* are already defined in Chap. 3 (Sect. 3.3.4.2). Thus, only the *trace share ratio* is defined below:

Trace share ratio: This is an indicator of the message delivery efficiency of the proposed routing protocol. The metric is defined as:

$$\text{Trace share ratio}\,(\%) = \frac{\text{Number of nodes received trajectory traces within a fixed time duration}}{\text{Total number of nodes in the network}} \times 100\%$$

The other three metrics such as *overhead ratio, average delivery latency* and *average energy consumption* are already defined in Chap. 3 (Sect. 3.3.4.2).

5.4.3.3 Results and Discussion

In order to evaluate the performance of the Trace Route routing protocol, four sets of experiments are conducted for a simulation interval of 2 h with respect to the design parameters and network parameter.

In the first set of experiments, the efficacy of all the competing routing protocols including the Trace Route in intermittent sharing of traces is estimated. The results are plotted in Fig. 5.5a. We observe from the results that the plots, in general, rise up to a certain time and then attain a steady state. To be more specific, after the entire simulation period, the *trace share ratio* is 95.2% for the Trace Route protocol, whereas this value is 73.52%, 75.64%, 94.57%, 93.71% and 94.28% for EASDM, SDM, MaxProp, PRoPHET and Epidemic, respectively.

We observe from the above comparative results that the Trace Route shows significant improvement in terms of trace share ratio over both the multicast routing protocols EASDM and SDM. Precisely, the improvement is approximately 29.48% and

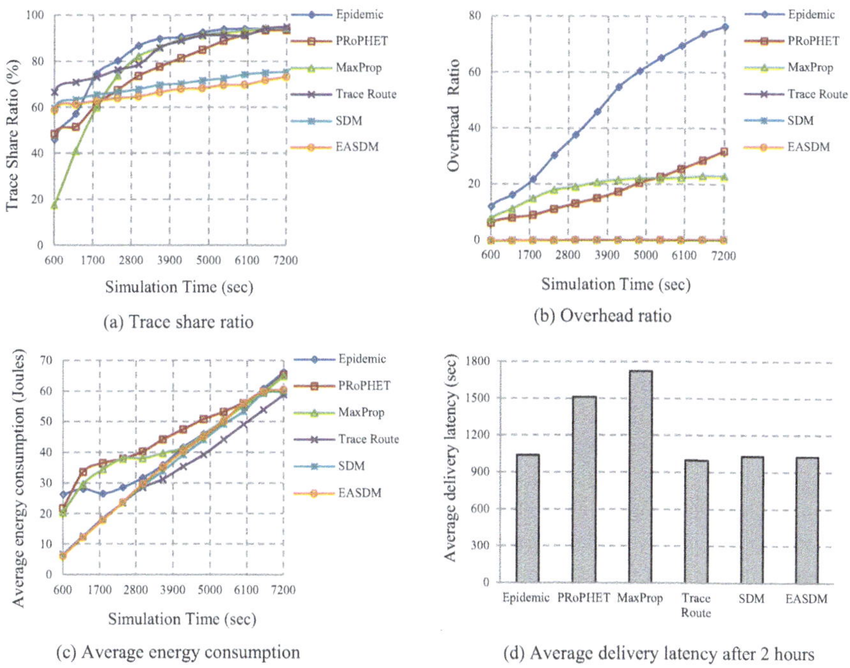

Fig. 5.5 Comparative performance of Trace Route protocol

25.72% in EASDM and SDM, respectively. However, this improvement is comparatively less than the unicast routing protocols. We also observe that subsequent to the creation of traces, the Trace Route distributes the same among more than 95% nodes in 2 h. This establishes the protocol's suitability in distributing the trajectory traces.

The second set of experiments measures the *overhead ratio* of all the competing routing protocols including the Trace Route over time. The results are plotted in Fig. 5.5b. We observe from the results that for the entire period of simulation, the *overhead ratio* is very marginal for all the multicast protocols such as SDM, EASDM and Trace Route. This implies that all the multicast protocols need very less message replication. However, for all the unicast protocols such as Epidemic, PRoPHET and MaxProp the *overhead ratio* is not so insignificant. Precisely, at the end of the entire simulation time, the *overhead ratio* is 0.57, 0.58, 0.6, 23.07, 32.04 and 76.61 for SDM, EASDM, Trace Route, MaxProp, PRoPHET and Epidemic, respectively. The reason for this significantly better performance of multicast protocols over unicast protocols is in unicast protocols delivering a single message to multiple nodes need the original message replication and the subsequent delivery of the replicas using unicast communication.

The average energy consumption of a node in delivering a message toward its target nodes using all the competing routing protocols including the Trace Route is measured. Figure 5.5c plots the results over time. We observe from the plots that the energy consumption rises over time for each of the protocols. We also observe that at the end of the entire simulation time, the Trace Route consumes 59.11 J, whereas this value is 60.74, 60.06, 65.19, 65.58 and 66.45 J for EASDM, SDM, MaxProp, PRoPHET and Epidemic, respectively. This implies that the Trace Route consumes the least energy.

The *average delivery latency* in delivering a message from its origin to the destinations is measured through the fourth set of experiments. The results are plotted in Fig. 5.5d for all the competing protocols. We observe from the results that at the end of the entire simulation period the *average delivery latency* of the Trace Route is 995.24 s, whereas this value is 1025.93 s, 1029.71 s, 1721.64 s, 1511.73 s and 1037.7 s for EASDM, SDM, MaxProp, PRoPHET and Epidemic, respectively. Thus, the Trace Route shows its superiority in terms of the network metric *average delivery latency* over all the competing routing protocols.

Summarily, the Trace Route shows its dominance primarily with respect to *trace share ratio*, one of our designated design goals, while requiring the least *average energy consumption*. It also shows its superiority in maintaining *average delivery latency* which is the network metric. The Trace Route exhibits better performance in the *overhead ratio* over all the protocols except EASDM and SDM which again lacks in performance with respect to *trace share ratio* and *average delivery* latency which are very important factors in distributing trajectory traces among the nodes of the network. Thus, the Trace Route shows its suitability in implementing group-based intermittent trace sharing in the post-disaster application domain.

The above-mentioned protocols such as Ge-PRoPHET and Trace Route are mainly designed to accommodate multicast communication. Thus, the evaluation of the protocols through simulation necessitates a simulator capable of supporting multicasting in opportunistic networks. The ONE simulator is extensively used by the research fraternity since it provides partial support for opportunistic network simulation. The existing version of the simulator (version 1.6.0) can only handle unicast communication of messages having textual content. Although dissemination among a collection of the nodes can be achieved through ONE simulator by forwarding single unicast messages to each node, the technique causes elevated network and storage overhead. In application domains such as post-disaster scenarios, the node connectivity, available bandwidth and storage are generally limited. Hence, disseminating separate unicast messages as a replacement of group communication proves infeasible in such scenarios due to high overhead. In addition to that, heterogeneous file formats need to be exchanged in such scenarios. Heterogeneity of file formats concerning information exchange involves sharing of multimedia content (images, audio and video, etc.) in addition to the textual content over opportunistic networks. Thus effective implementations of multicast and broadcast communications, as well as propagation of multimedia content are essential in present opportunistic network research. Hence, in the next section, the enhancement of ONE simulator to facilitate multicast and broadcast communication for multimedia data is illustrated.

5.5 Implementation of Multicasting and Broadcasting in ONE Simulator

Simulations are critical for studying the characteristics of opportunistic network routing and application protocols. One major limitation of opportunistic network research is the insufficiency of suitable simulators. Existing simulators for MANETs (e.g. *ns2* [28], *ns3* [29] and *OMNeT++* [30]) and opportunistic network (e.g. *dtnsim* [31] and *dtnsim2* [32]) does not offer full support for opportunistic network simulation. Hence, there is a necessity for a simulator that can realistically simulate the opportunistic network scenarios. ONE simulator [21] incorporates all the essential features of opportunistic networks. As opposed to the aforementioned benefits, ONE simulator only supports the propagation of textual messages using unicast communication. In this section, the development of the multicasting and broadcasting of multimedia content in ONE simulator version 1.5.1 RC-2 [24] is illustrated.

5.5.1 Overview of ONE Simulator

The ONE is an agent-based discrete event simulator written in Java. Figure 5.6 portrays the architecture of ONE simulator. The entire simulator is constructed using six modules such as movement model, event generator, simulation engine, routing, visualization and report.

Fig. 5.6 ONE simulator modules and their interactions

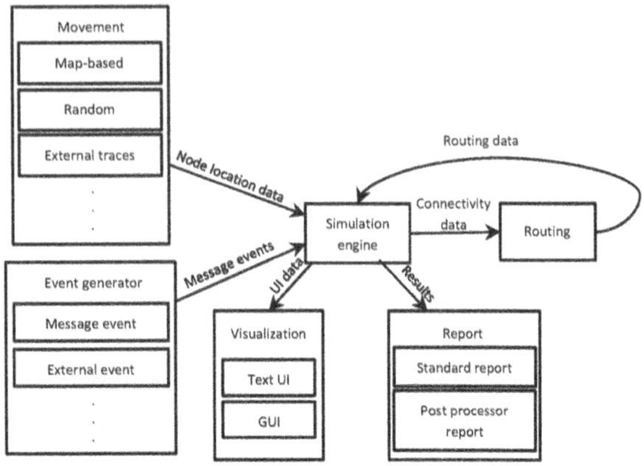

The movement model module comprises ten classes representing various mobility models that control node movement during the simulation. The message events are generated through the event generator module. The diverse parameters required for message creation such as creation time, message size, destination range, the frequency of message creation, etc. are managed by the event generator module. The routing module is constructed using 14 classes related to various opportunistic network routing protocols. The primary function of this module is message forwarding. The movement modeling and routing simulation are exhibited in the simulator's visualization module. The report module collects the simulation results for further study.

The modification details regarding multicasting, broadcasting and handling of multimedia data are discussed in the subsequent sections.

5.5.2 Implementing Multicasting and Broadcasting

In this section, the modifications of ONE simulator to incorporate multicasting and broadcasting are discussed. The modification to incorporate multicasting and broadcasting is done in six classes. Out of these, three classes (*MulticastEventGenerator*, *MessageEventGenerator* and *MessageCreateEvent*) are connected to the event generator module. Two classes (*MessageRouter* and *ActiveRouter*) are connected to the routing module. The *Message* class is linked with the simulation engine module. The class diagram of the proposed modification is illustrated in Fig. 5.7. The modification consists of the steps such as developing a multicast event generator, modifying the routing module and modifying the settings file.

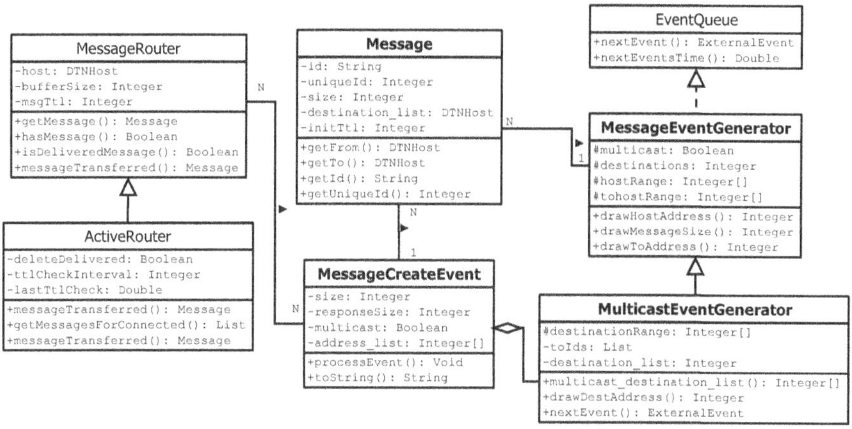

Fig. 5.7 Class diagram of multicast/broadcast implementation

5.5.2.1 Developing the Multicast Event Generator

As discussed earlier, the ONE simulator typically forwards messages in unicast mode. Hence, there is a necessity to develop a new message event generator with multiple destination-oriented message creation capabilities. Figure 5.8 illustrates the modified *MulticastEventGenerator* constructor. A new parameter, *MULTICAST_DEST_RANGE* is added in the constructor which takes care of the multicast/broadcast destination addresses. Further, a couple of methods such as *drawDestAddress()* and *multicast_destination_list()* are also developed for selecting the destination range from the settings file and authentication of the address range.

```
public MulticastEventGenerator(Settings s) {
                super(s);
                this.destinationRange = s.getCsvInts(MULTICAST_DEST_RANGE, 2);
                this.toIds = new ArrayList<Integer>();
                if (toHostRange == null) {
                        throw new SettingsError("Destination host (" + TO_HOST_RANGE_S + ") must be
defined");
                }

                for (int i = toHostRange[0]; i < toHostRange[1]; i++)
                        toIds.add(i);
                if ((destinationRange[1]-destinationRange[0]) != destinations){
                        throw new SettingsError("Destination address range (" + MULTICAST_DEST_RANGE +
") and number of destination (" + MULTICAST_DEST_S + ") must be identical");
                }
                        destination_list = new int[]{destinationRange[0], destinationRange[1]};
                Collections.shuffle(toIds, rng);
        }
```

Fig. 5.8 *MulticastEventGenerator* constructor

The *MessageCreateEvent* class is intended for generating message events using the parameters provided by the user. Two additional parameters (*multicast* and *destination_list*) are included in the class constructor to facilitate multicasting/broadcasting. Figure 5.9 demonstrates the alteration performed on the *MessageCreateEvent* constructor.

The *Message* class in the simulation engine typically generates messages using diverse attributes provided by the users using the settings file. The multicasting/broadcasting necessitates the provision to accommodate multiple addresses in the destination address field of the message. Hence, the Message class constructor is also modified by incorporating a destination address list in it. Figure 5.10 illustrates the altered *Message* constructor.

```
public MessageCreateEvent(int from, int to, String id, int size, int responseSize, double time, boolean
multicast,int destination_list[]) {
          super(from,to, id, time);
          this.size = size;
          this.responseSize = responseSize;
          this.multicast = multicast;
this.destination_list = destination_list; }
```

Fig. 5.9 Modified *MessageCreateEvent* constructor

```
public Message(DTNHost from, DTNHost to, String id, int size,ArrayList<DTNHost>
destination_list) {
this.from = from;
this.to = to;
this.id = id;
this.size = size;
this.path = new ArrayList<DTNHost>();
this.uniqueId = nextUniqueId;
this.timeCreated = SimClock.getTime();
this.timeReceived = this.timeCreated;
this.initTtl = INFINITE_TTL;
this.responseSize = 0;
this.requestMsg = null;
this.properties = null;
this.appID = null;
this.address_list = destination_list;
Message.nextUniqueId++;
addNodeOnPath(from);}
```

Fig. 5.10 Modified *Message* constructor

5.5.2.2 Modifying the Routing Module

The ONE simulator conventionally supports unicast communication. Therefore, forwarding the messages to only one destination is possible. Upon reaching the destination the messages are not spread further. However, in case of multicasting/broadcasting, the message spreading is not finished till all the destinations receive the message. Hence, the routing module needs to be modified. The routing module contains a couple of abstract classes (*MessageRouter* and *ActiveRouter*). The *MessageRouter* class comprises a method called *messageTransferred()* which is accountable for message termination upon reaching the required destinations.

A destination list is introduced in the *messageTransferred()* method which eliminates the possibility of message termination upon reaching a solitary destination.

The *ActiveRouter* class contains a method called *getMessagesForConnected()* accountable for producing a list of message-connections tuples. Modifications are also done in this method to facilitate multicasting/broadcasting.

5.5.2.3 Modifying the Settings File

Figure 5.11 presents the picture of a sample settings file for multicasting/broadcasting in the ONE simulator. This file retains the values of diverse parameters.

Three parameters (*multicast*, *nrofDestination* and *destinationRange*) are added to the settings file. The first one (*multicast*) is a Boolean parameter having either true or false values. If the value of *multicast* is false the simulator performs unicast communication. The *nrofDestination* is required to recognize the quantity of multicast destinations. If the *nrofDestination* comprises all the nodes in the simulation environment the simulator simply broadcasts messages. The *destinationRange* parameter specifies the collection of the destination addresses.

```
Events2.class =MulticastEventGenerator
Events2.interval = 1
Events2.size = 50k, 250k
Events2.hosts = 2,2
Events2.prefix = INV_MW2_
Events2.multicast = true
Events2.nrofDestination = 23
Events2.destinationRange = 1,24
```

Fig. 5.11 Modified settings file

5.5.3 *Processing Multimedia Data*

The modification for multimedia content (image) handling in the ONE simulator is described in this section. The simulator only supports textual message dissemination. In order to mitigate this shortcoming, the simulator is modified to handle image files by converting them to text files. The transformation is achieved by converting an image into grayscale and preserving the grayscale luminance values in a text file. To accomplish this integration of the Open Source Computer Vision Library version 2.4.13 (*OpenCV*) is necessary. *OpenCV* is an open-source computer vision and machine learning software library written in C++.

The multimedia data processing is accomplished by three steps such as extracting text from the image, modifying the simulation engine module and reconstructing the image at the destinations. Figure 5.12 shows the class diagram for handling multimedia data in ONE simulator. The modification needs to be done in six classes. Out of these six, two classes (*ActiveRouter* and *Reconstruct*) are connected with the routing module of the ONE simulator. Two classes (*DTNHost* and *Message*) are connected with the simulation engine module and two classes (*MessageCreateEvent* and *Extract*) are connected with the event generator module.

5.5.3.1 Extracting Text from Image

The technique for extracting text from the image files and creating messages from such textual data is described in this section. A class named *Extract* is developed for this purpose. This class introduces a method called *grayscale()* which is accountable

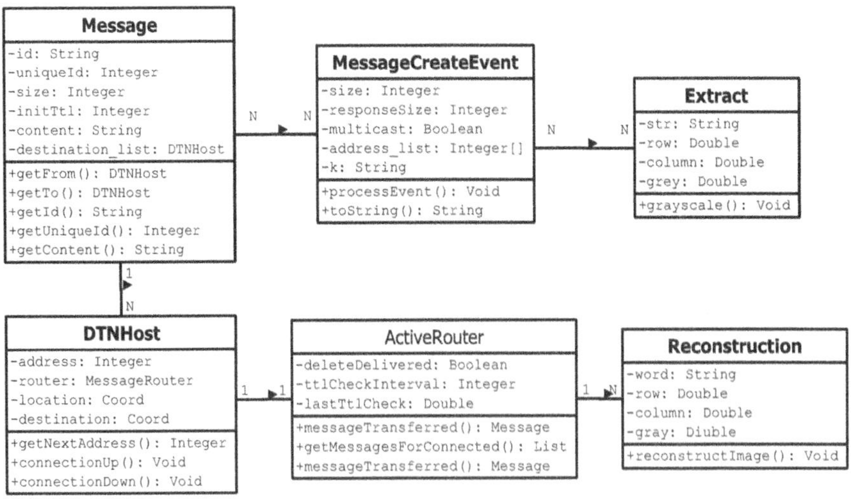

Fig. 5.12 Class diagram of processing multimedia data

```
public static void grayscale()throws IOException{
    System.loadLibrary(Core.NATIVE_LIBRARY_NAME);
    Mat frame=new Mat();int i=0,j=0;
    String strr,str,strrr;double row,col,gray;
    String k;
    strr="input/content/image.jpg";
    frame = Highgui.imread(strr);
    Size size =frame.size();
    row=size.height;col=size.width;
    FileOutputStream fos = new FileOutputStream("input/content/image_data.txt");
    OutputStreamWriter osw=new OutputStreamWriter(fos);
    BufferedWriter bw=new BufferedWriter(osw);
    for (i = 0; i <row; i++) {
      for (j = 0; j <col; j++) {
        double[] data = frame.get(i, j);
        gray=(data[0]+data[1]+data[2])/3;
        k=Double.toString(gray);
        bw.write(k+" ");
      }
    }
    bw.close();
    osw.close();
    fos.close();
}
```

Fig. 5.13 The *grayscale()* method of Extract class

for the extraction of grayscale luminance values from red, green and blue (RGB) luminance values and preserving such data in a text file. The conversion of RGB to grayscale is accomplished by computing the arithmetic mean of the RGB luminance values for each pixel. Figure 5.13 illustrates the *grayscale()* method. According to the figure, the *gray* variable preserves the arithmetic mean of the RGB luminance for each pixel and stores them in the *image_data.txt* file.

The *MessageCreateEvent* constructor is also altered to enable the construction of messages comprising data from the *image_data.txt* file. Figure 5.14 portrays the altered *MessageCreateEvent* constructor. The *MessageCreateEvent* class generates text files extracted from images for each message producing nodes. The substances of those text files are added to the corresponding message body during the construction of the messages.

5.5.3.2 Modifying the Simulation Engine Module

The alterations performed in the simulation engine module to accommodate the dissemination of the text files in the ONE simulator are discussed in this section.

```
public MessageCreateEvent(int from, int to, String id, int size, int responseSize, double time,
boolean multicast, int destination_list[]) {
        super(from,to, id, time);
        this.size = size;
        this.responseSize = responseSize;
      this.multicast = multicast;
      this.destination_list = destination_list;
        String k;
        if(id.startsWith("M")){
                try {
                        extract.grayscale();
                        s =  new Scanner(new File("input/content/image_data.txt"));
                        content = s.nextLine();
                }
                catch(Exception e){
                        System.out.println("ERROR");
                }
                try{
                        k=Integer.toString(from);
                        FileOutputStream fos = new FileOutputStream("Game/"+k+"/image_data.txt");
                        OutputStreamWriter osw=new OutputStreamWriter(fos);
                        BufferedWriter bw=new BufferedWriter(osw);
                        bw.write(content);
                        bw.close();
                }
                catch(Exception e){
                        System.out.println("ERROR");
                }
        }
}
```

Fig. 5.14 Modified *MessageCreateEvent* constructor

The class called *DTNHost* typically creates fresh nodes in the ONE simulator. The constructor of *DTNHost* is modified to enable the formation of folders bearing the names of the opportunistic network nodes so that the forwarded messages can be tracked. The folders bearing the labels of the message originator nodes and relay nodes store the text files produced from images. The folders bearing the labels of destination nodes contain text files and reproduced image files. In addition to that, the constructor of the *Message* class is also altered so that the data from the text files can be appended to the message body.

Figure 5.15 portrays the modified *Message* constructor. The figure elucidates the introduction of a new string variable *content* in the constructor. This variable retains

```
public Message(DTNHost from, DTNHost to, String id, int size,ArrayList<DTNHost> destination_list,
String content){
                this.content=content;
                this.address_list = destination_list;
                Message.nextUniqueId++;
                addNodeOnPath(from);
}
```

Fig. 5.15 Modified *Message* constructor

the textual data produced from the file *image_data.txt*. A method *getContent()* is also developed in the *Message* class to retrieve the message content.

5.5.3.3 Reconstructing Image File at Destinations

The technique of reproducing the image files at the destinations is presented in this section. Here, the reproductions of images are performed at destinations. Hence, the reproductions mechanism is included in the routing module of the ONE simulator.

The *ActiveRouter* class of the routing module comprises the *messageTransferred()* method accountable for eliminating messages upon reaching the intended destinations. Figure 5.16 specifies the alteration done in the *messageTransferred()* method the reproduction of the image file at the destination. The figure shows that *reconstructImage()* method is invoked once the message reaches the destination.

The *Reconstruction* class is developed containing the *reconstructImage()* method for that purpose. The method receives the data from the text file and reproduces the equivalent image file. The reproduced image file is preserved in the folders bearing the names of the destination nodes.

```
public Message messageTransferred(String id, DTNHost from) {
    Message m = super.messageTransferred(id, from);
    String data=m.content;
    String to=getHost().toString();
    System.out.println(to);
    try{
        FileOutputStream fos = new FileOutputStream("Game/"+to+"/image_data.txt");
        OutputStreamWriter osw=new OutputStreamWriter(fos);
        BufferedWriter bw=new BufferedWriter(osw);
        bw.write(data);
        bw.close();
        if(getHost()== m.getTo()){
            System.out.println("Reconstruction Started");
            reconstruction.reconstructImage(to);
        }
    }
    catch(Exception e){
        System.out.println("Message delivery error.");
    }
    if (m.getTo() == getHost() && m.getResponseSize() > 0) {
        Message res = new Message(this.getHost(),m.getFrom(), RESPONSE_PREFIX+m.getId(),
m.getResponseSize(), m.content);
                        this.createNewMessage(res);
        this.getMessage(RESPONSE_PREFIX+m.getId()).setRequest(m);
        }
    return m;
}
```

Fig. 5.16 Modified *messageTransferred()* method

5.5.4 Performance Evaluation

This section provides the performance estimation of multicasting/broadcasting of text and image data within ONE simulator through two sets of experiments.

5.5.4.1 Simulation Setup

The simulation setup is similar to the setup as mentioned in Chap. 3 (Sect. 3.3.4.1). On the map of the affected area of Kathmandu, as demonstrated in Fig. 3.9 of Chap. 3, seven camps are set up within an area of 3.5 km^2 around the Patan area. The other parameters such as the number of *volunteer nodes*, *ferry nodes*, mobility models and network architecture are similar to Sect. 5.2.5.1. However, unlike Sect. 5.2.5.1, here the number of camp nodes is increased to seven. Moreover, as opposed to Sect. 5.2.5.1, in this section, the mode of communication for simulation is chosen as group-based multicasting.

The simulation environment is similar to the environment described in Sect. 5.3.3.1. Further, in the first set of experiments, we consider textual messages are originated intermittently from all the camp nodes daily. We analyze the performance using all the unicast, multicast and broadcast communication modes. We consider that in unicast communication, out of seven generated messages, one of each is sent to the seven volunteer nodes chosen randomly. On the other hand, in the multicast mode of communication, each of the seven generated messages is sent to 22 designated nodes.

In the second set of experiments, we consider multimedia data (image) are originated intermittently from seven volunteer nodes chosen randomly at 2.5 h interval. Thus, in a simulation period of 8 h, 21 messages are originated. Out of these 21 messages, in unicast communication, three messages are sent to each of the seven destinations, whereas in multicast communication, each of such 21 messages is sent to 15 destinations. All these destinations are randomly chosen volunteer nodes.

5.5.4.2 Results and Discussion

This section estimates the message delivery over time through the first set of experiments in all the three modes of communication. The comparative results are plotted in Fig. 5.17. We observe from the plots that compared to unicast communication, message delivery is significantly higher at any instance of time in multicast/broadcast communication. We also observe that in multicast/broadcast communication, the number of messages delivered is greater than the number of messages generated. This definitely confirms that each of the messages has been carried to multiple destination nodes implying success of realizing both the multicasting and broadcasting in ONE.

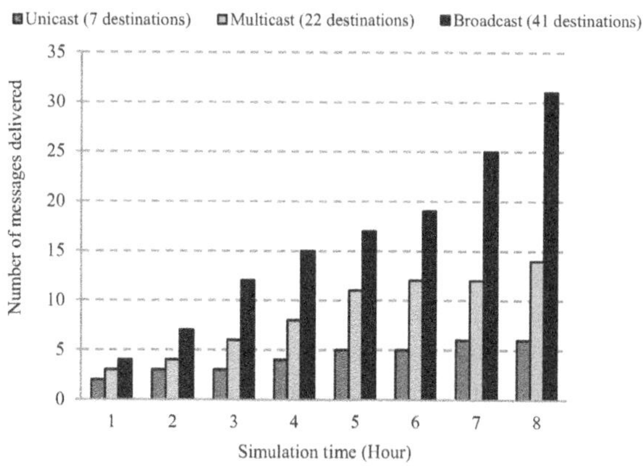

Fig. 5.17 Comparative performance of three modes of communication

In support of our claim of successful implementation of multicast/broadcast communication, we also provide the message stats reports of ONE in Fig. 5.18. These reports show all the relevant details of both the text and image data dissemination in the unicast/multicast/broadcast mode at the end of the entire simulation

```
sim_time: 28800.1000
created: 7
started: 105
relayed: 92
aborted: 7
dropped: 0
removed: 0
delivered: 6
latency_avg: 6218.4513
latency_med: 4716.6000
hopcount_avg: 2.4882
hopcount_med: 2
```

Unicast mode

```
sim_time: 28800.0000
created: 7
started: 293
relayed: 254
aborted: 23
dropped: 0
removed: 0
delivered: 14
latency_avg: 2342.7524
latency_med: 1396.1000
hopcount_avg: 2.2451
hopcount_med: 2
```

Multicast mode

```
sim_time: 28800.0000
created: 7
started: 313
relayed: 277
aborted: 36
dropped: 0
removed: 0
delivered: 31
latency_avg: 2355.7386
latency_med: 1436.1000
hopcount_avg: 2.2362
hopcount_med: 2
```

Broadcast mode

(a) Textual

```
sim_time: 28800.0000
created: 21
started: 424
relayed: 679
aborted: 56
dropped: 0
removed: 0
delivered: 16
latency_avg: 3245.2374
latency_med: 2987.5000
hopcount_avg: 1.4823
hopcount_med: 1
```

Unicast mode

```
sim_time: 28800.0000
created: 21
started: 306
relayed: 299
aborted: 7
dropped: 0
removed: 0
delivered: 131
latency_avg: 4280.1154
latency_med: 3115.6000
hopcount_avg: 1.4615
hopcount_med: 1
```

Multicast mode

```
sim_time: 28800.0000
created: 21
started: 352
relayed: 303
aborted: 15
dropped: 0
removed: 0
delivered: 459
latency_avg: 3678.3072
latency_med: 3274.4000
hopcount_avg: 1.3829
hopcount_med: 1
```

Broadcast mode

(b) Multimedia

Fig. 5.18 Message stats report after 8 h of simulation

period. Here also, we notice that the number of both the text and image messages delivered is greater than the number of messages originated in multicast/broadcast mode. This conforms to the results of Fig. 5.17.

Finally, in Fig. 5.19, we present some of the sample images used in the simulation, both in original and reconstructed forms. As the original images are converted into grayscale (Sect. 5.5.3) at the sender side, the figure presents grayscale images both at the sender as well as the receiver side as original and reconstructed images, respectively. Hence, from Figs. 5.17, 5.18 and 5.19, we establish our claim of successful realization of multicasting/broadcasting of image (multimedia) through the modified ONE simulator developed by us. The modified ONE is uploaded in Github at https://github.com/suman-iiest/ModifiedONE_1.5.1RC2.

Fig. 5.19 Snapshots of original and reconstructed image files

5.6 Conclusion

This chapter presents three different opportunistic network routing protocols designed to fit into the application of post-disaster communication. Out of these three routing protocols, one is intended for unicast communication and the other two are intended for multicast communication. Each of these protocols has its own merits and demerits. For instance, the Pen-PRoPHET routing protocol delivers messages containing critical information to a destination (unicast) on a priority basis, but this protocol lacks in spreading messages to a group of pre-defined destinations (multicast). Similarly, the Ge-PRoPHET routing protocol delivers messages to a group of pre-defined destinations based on their encounter patterns with other nodes within a strict time constraint. But, this protocol is unable to share messages among all the nodes (broadcast) in a specific disaster-affected area. Finally, the Trace Route protocol shares messages among all the nodes (broadcast) in a network.

However, the simulation results indicate that on average all the proposed routing protocols exhibit around 94% *delivery ratio* along with low energy consumption and reduced delay during message transfer compared to conventional opportunistic network routing protocols. Hence, the aforesaid routing protocols with suitable underlying opportunistic network architecture can be successfully employed for providing post-disaster management services.

References

1. Vogels, W., van Renesse, R., Birman, K.: The power of epidemics. ACM SIGCOMM Comput. Commun. Rev. **33**, 131–135 (2003). https://doi.org/10.1145/774763.774784
2. Spyropoulos, T., Psounis, K., Raghavendra, C.S.: Spray and wait: an efficient routing scheme for intermittently connected mobile networks. In: Proceedings of the 2005 ACM SIGCOMM Workshop on Delay-Tolerant Networking—WDTN '05, pp. 252–259 (2005). https://doi.org/10.1145/1080139.1080143
3. Lindgren, A., Doria, A., Schelén, O.: Probabilistic routing in intermittently connected networks. ACM SIGMOBILE Mob. Comput. Commun. Rev. **7**, 19 (2003). https://doi.org/10.1145/961268.961272
4. Grasic, S., Davies, E., Lindgren, A., Doria, A.: The evolution of a DTN routing protocol—PRoPHETv2. In: Proceedings of the 6th ACM Workshop on Challenged Networks—CHANTS '11, p. 27. ACM Press, New York, NY, USA (2011)
5. Burgess, J., Gallagher, B., Jensen, D., Levine, B.N.: MaxProp: routing for vehicle-based disruption-tolerant networks. In: Proceedings of the IEEE INFOCOM 2006. 25TH IEEE International Conference on Computer Communications, pp. 1–11. IEEE (2006)
6. Balasubramanian, A., Levine, B., Venkataramani, A., Balasubramanian, A., Levine, B., Venkataramani, A.: DTN routing as a resource allocation problem. In: Proceedings of the 2007 Conference on Applications, Technologies, Architectures, and Protocols for Computer Communications—SIGCOMM '07, p. 373. ACM Press, New York, NY, USA (2007)
7. Hui, P., Crowcroft, J., Yoneki, E.: BUBBLE Rap: social-based forwarding in delay-tolerant networks. IEEE Trans. Mob. Comput. **10**, 1576–1589 (2011). https://doi.org/10.1109/TMC.2010.246
8. Joe, I., Kim, S.-B.: A Message Priority Routing Protocol for Delay Tolerant Networks (DTN) in Disaster Areas, 13 Dec 2010

9. Liu, X., Chen, Y., Li, C., Taylor, W., Moore, J.H.: Message prioritization of epidemic forwarding in delay-tolerant networks. In: 2014 International Conference on Computing, Networking and Communications (ICNC), pp. 573–577. IEEE (2014)
10. Chuah, M.C., Yang, P.: Context-aware multicast routing scheme for Disruption Tolerant Networks. Int. J. Ad Hoc Ubiquitous Comput. **4**, 269 (2009). https://doi.org/10.1504/IJAHUC.2009.027477
11. Lee, U., Oh, S.Y., Lee, K.W., Gerla, M.: RelayCast: scalable multicast routing in delay tolerant networks. In: 2008 IEEE International Conference on Network Protocols, pp. 218–227. IEEE (2008)
12. Chuah, M.C., Choo, M.: Social network aided multicast delivery scheme for human contact-based networks. In: Proceedings of the 1st Annual Workshop on Simplifying Complex Network for Practitioners—SIMPLEX '09, pp. 1–6. ACM Press, New York, NY, USA (2009)
13. Gao, W., Li, Q., Zhao, B., Cao, G.: Social-aware multicast in disruption-tolerant networks. IEEE/ACM Trans. Netw. **20**, 1553–1566 (2012). https://doi.org/10.1109/TNET.2012.2183643
14. Galluccio, L., Lorenzo, B., Glisic, S.: Sociality-aided new adaptive infection recovery schemes for multicast DTNs. IEEE Trans. Veh. Technol. **65**, 3360–3376 (2016). https://doi.org/10.1109/TVT.2015.2450202
15. Roy, A., Bose, S., Acharya, T., DasBit, S.: Social-based energy-aware multicasting in delay tolerant networks. J. Netw. Comput. Appl. **87**, 169–184 (2017). https://doi.org/10.1016/J.JNCA.2017.03.020
16. Sobin, C.C., Raychoudhury, V., Marfia, G., Singla, A.: A survey of routing and data dissemination in Delay Tolerant Networks. J. Netw. Comput. Appl. **67**, 128–146 (2016). https://doi.org/10.1016/J.JNCA.2016.01.002
17. Grossglauser, M., Tse, D.N.C.: Mobility increases the capacity of ad hoc wireless networks. IEEE/ACM Trans. Netw. **10**, 477–486 (2002). https://doi.org/10.1109/TNET.2002.801403
18. Bhattacharjee, S., Basu, S., Roy, S., Das Bit, S.: Best-effort delivery of emergency messages in post-disaster scenario with content-based filtering and priority-enhanced PRoPHET over DTN. In: 2016 8th International Conference on Communication Systems and Networks, COMSNETS 2016 (2016)
19. Shi, H., Liu, Y.: Naïve Bayes vs. Support Vector Machine: Resilience to Missing Data
20. Hall, M., Frank, E., Holmes, G., Pfahringer, B., Reutemann, P., Witten, I.H.: The WEKA data mining software. ACM SIGKDD Explor. Newsl. **11**, 10 (2009). https://doi.org/10.1145/1656274.1656278
21. Keränen, A., Ott, J., Kärkkäinen, T.: The ONE simulator for DTN protocol evaluation. In: Proceedings of the Second International ICST Conference on Simulation Tools and Techniques, p. 55. ICST (2009)
22. Bhattacharjee, S., Roy, S., DasBit, S.: DPDRM: a decentralized post-disaster resource management scheme using energy efficient smart phone based DTN. J. Netw. Comput. Appl. **111** (2018). https://doi.org/10.1016/j.jnca.2018.03.007
23. Basu, S., Bhattacharjee, S., Roy, S., Bandyopadhyay, S.: SAGE-PRoPHET: a security aided and group encounter based PRoPHET routing protocol for dissemination of post disaster situational data. In: ACM International Conference Proceeding Series (2015)
24. Bhattacharjee, S., Chatterjee, R., Pal, T., DasBit, S.: Implementing multicasting and broadcasting of multimedia data in ONE simulator. In: Proceedings of the 10th EAI International Conference on Simulation Tools and Techniques—SIMUTOOLS '17, pp. 48–56 (2017). https://doi.org/10.1145/3173519.3173540
25. Cao, Y., Sun, Z.: Routing in delay/disruption tolerant networks: a taxonomy, survey and challenges. IEEE Commun. Surv. Tutor. **15**, 654–677 (2013). https://doi.org/10.1109/SURV.2012.042512.00053
26. Bhattacharjee, S., Roy, S., Bit, S.D.: Post-disaster map builder: crowdsensed digital pedestrian map construction of the disaster affected areas through smartphone based DTN. Comput. Commun. (2018). https://doi.org/10.1016/J.COMCOM.2018.11.010
27. Ben-Zur, L.: Trepn Power Profiler—Qualcomm Developer Network. https://developer.qualcomm.com/software/trepn-power-profiler

28. The Network Simulator—ns-2. https://www.isi.edu/nsnam/ns/
29. The Network Simulator—ns-3. https://www.nsnam.org/
30. Varga, A., Hornig, R.: An overview of the OMNeT++ simulation environment. In: Proceedings of the 1st International Conference on Simulation Tools and Techniques for Communications, Networks and Systems & Workshops, vol. 60 (2008)
31. Jain, S., Fall, K., Patra, R., Jain, S., Fall, K., Patra, R.: Routing in a delay tolerant network. In: Proceedings of the 2004 Conference on Applications, Technologies, Architectures, and Protocols for Computer Communications—SIGCOMM '04, p. 145. ACM Press, New York, NY, USA (2004)
32. DTNSim2 Delay-tolerant Network Simulator. https://ex001slvr.wordpress.com/category/dtnsim2/

Chapter 6
Conclusion and Future Scope

With the increasing number of disasters across the world in recent times, it has become mandatory for each of the disaster-prone countries to equip themselves with appropriate measures to cope with the challenges of providing post-disaster services. Some of such challenges are incapacitated communication infrastructure, unstable power supply and inaccessible road networks. Out of these challenges, the destruction of road networks, especially in developing countries, acts as a major hindrance to effective disaster management. To be more specific, the success of a disaster response operation generally depends on the speed of evacuation and transportation of an adequate amount of relief materials at the right time to the disaster-affected areas. Hence, map-based navigation support is a primary necessity for disaster response operations. This monograph provides the details of developing a digital route map construction system over opportunistic networks. The monograph also provides the solution of the two other important post-disaster management services such as situational awareness and resource allocation. Both of these services are invariably dependent on the existence of navigation support. In order to offer such services, the other challenges that need to address are the problems of incapacitated communication infrastructure and unstable power supply. This monograph addresses such challenges and develops the solutions for the said post-disaster management services.

The overall analysis of the research works discussed in this monograph and the highlights of the key contributions are summarized in this final chapter. In addition to that, the chapter provides future research directions in the area of offering reliable post-disaster management services in absence of traditional communication infrastructure. The chapter ends with concluding remarks.

S. Bhattacharjee et al., *Post-disaster Navigation and Allied Services over Opportunistic Networks*, Smart Innovation, Systems and Technologies 228,
https://doi.org/10.1007/978-981-16-1240-4_6

6.1 Review of Research

This section summarizes the works presented in this monograph. As mentioned earlier, map-based navigation service becomes a critical prerequisite for post-disaster response operations. Such service guides volunteers to reach their destinations, helps victims to locate the shelters, and support disaster response agencies in forwarding their resources in the disaster-struck areas. However, the existing routes for reaching different destinations alter significantly in the disaster-struck areas in disaster aftermath due to the road blockage/destruction caused by disasters. Hence, the existing route maps of the disaster-struck areas become outdated. Thus, in Chap. 2, a smartphone-based digital route map construction system (Post-disaster Map Builder) is discussed. The Post-disaster Map Builder system constructs route maps of the affected areas using mobile handheld devices. Here, GPS traces accumulated by the volunteers are exchanged with the other volunteers of the disaster-struck areas through opportunistic networks. Route maps of the affected areas are progressively constructed in the smartphones of the volunteers by merging the traces over time. The system is evaluated using simulation and testbed implementation and the results illustrate the capability of the system for constructing route maps of the affected areas with high accuracy within an acceptable delay.

In Chap. 3, two other post-disaster management services such as situational awareness and resource management are developed through the Disaster Messenger app and DPDRM scheme, respectively. The Disaster Messenger facilitates the dissemination of situational information in absence of network infrastructure in unicast, multicast and broadcast mode. It harnesses the potential of Wi-Fi direct to offer the functionalities like opportunistic networks and constructs fresh threads for transmission and reception. Hence transmission and reception of data to/from numerous sources can be accomplished in parallel. It also outperforms well-known Android-based file sharing applications SHAREit and BitTorrent Sync in terms of energy efficiency and data transfer rate in unicast mode.

The DPDRM is designed to automate post-disaster resource management. The scheme facilitates group-based sharing of resource inventories causing effective search and retrieval of queries within the desired time period. Here, resource inventories are shared among the group members by employing the publish-subscribe mechanism. Moreover, the queries are forwarded to the group members depending on the likelihood of meetings. This mechanism controls the unintentional duplication of the query. The experimental results signify the proficiency of the DPDRM scheme in the disaster aftermath. For instance, the DPDRM scheme successfully delivers around 88% response to the query originator nodes within 8 h from the initiation of the query and consumes around 28% of the stored energy.

In order to offer the services mentioned in Chaps. 2 and 3, the existence of a post-disaster opportunistic network framework becomes crucial. Such a framework comprises two entities such as opportunistic network architecture and energy-efficient post-disaster routing protocols.

In Chap. 4, reliable and energy-efficient opportunistic network architecture is designed to support post-disaster communication. The opportunistic network architecture includes a *three-tier* network architecture involving mobile and static nodes. However, performance evaluation of the proposed *three-tier* architecture indicates that the message delivery up to 63% can be attained by implementing the aforesaid network architecture, which is inadequate for post-disaster message transfer. In order to improve the performance of the opportunistic network further, the appropriate node deployment plan for the nodes in *tier-1* and *tier-2* proves beneficial. Hence, suitable node deployment plans for mobile and static nodes are discussed.

It has been observed that the introduction of an appropriate node deployment plan for both static and mobile nodes improves the message delivery up to 86% with average delivery latency of 631.23 s. The experimental results reveal that the *three-tier* network architecture along with a suitable node deployment plan does not entirely satisfy the desired requirements for post-disaster message transfer. This motivates us to develop suitable opportunistic network routing protocols for each of the three post-disaster management services discussed in this monograph toward improving the said performance further.

In Chap. 5, three different routing protocols supporting three different post-disaster management services are proposed. Out of these three routing protocols one is intended for unicast communication (Pen-PRoPHET) and the other two are intended for multicast communication. Each of these protocols has its own merits and demerits. For instance, the Pen-PRoPHET routing protocol delivers messages containing critical information to a destination (unicast) on a priority basis. Thus, it is suitable for situational awareness service. However, this protocol lacks in spreading messages to a group of pre-defined destinations (multicast). Similarly, the Ge-PRoPHET routing protocol delivers messages to a group of pre-defined destinations based on their encounter patterns with other nodes within a strict time constraint. This protocol is suitable for resource management service through periodic group-based multicasting of resource inventories and forwarding of queries. But, this protocol is unable to share messages among all the nodes (broadcast) in a specific disaster-affected area. Finally, the Trace Route protocol shares messages among all the nodes (broadcast) in a network. This protocol is suitable for map-based navigation service by sharing trajectory traces with all the nodes in a network.

However, the simulation results indicate that on average all the routing protocols described in Chap. 5 exhibit around 94% *delivery ratio* along with lowered energy consumption and reduced delay during the message transfer. Hence, the aforesaid routing protocols along with suitable communication architecture can be successfully employed for developing post-disaster management services.

6.2 Future Research Scope

This section gives directions toward the possible future works that can be carried out by expanding the works described in this monograph.

- The Post-disaster Map Builder system may be made more realistic from the following perspectives:
  - Cross-platform implementation of the Map Builder may be explored.
  - Automated shortest path navigation assistance for volunteers may be added in such Map Builder.
- The proposed DPDRM scheme may be improved to incorporate the following:
  - Instead of providing the recommended lists of camps in the entire network area for catering a resource request, the camps located only within the access of volunteers may be provided.
  - Inventory sharing module may be improved by reducing the delay of rendering synchronous resource inventory further.
- The Pen-PRoPHET routing protocol may be improved by activating the message prioritization module adaptively based on the average residual energy of the nodes and the degree of network connectivity. This mechanism will conserve energy and enhance network lifetime.

6.3 Concluding Remarks

This monograph explores the potential of opportunistic networks in providing map-based navigation support in post-disaster scenarios in absence of conventional communication infrastructure. Moreover, it addresses several challenges in post-disaster scenarios such as the lack of situational awareness and ad hoc allocation and distribution of resources. It also validates the significance of the proposed systems, schemes, architectures and protocols in disaster response operations. Finally, this monograph provides future directions toward improving the entire post-disaster services presented here and significant research challenges associated with such services.

Index

S. Bhattacharjee et al., *Post-disaster Navigation and Allied Services over Opportunistic Networks*, Smart Innovation, Systems and Technologies 228,
https://doi.org/10.1007/978-981-16-1240-4

The manufacturer's authorised representative in the EU is Springer Nature Customer Service Centre GmbH, Europaplatz 3, 69115 Heidelberg, Germany. If you have any concerns regarding our products, please contact ProductSafety@springernature.com

Printed and bound by CPI Group (UK) Ltd, Croydon, CR0 4YY
30/09/2026
02225307-0019